FUNGIBLE LIFE

D0733071

FUNGIBLE *LIFE*

EXPERIMENT IN THE ASIAN CITY OF LIFE

AIHWA ONG

Duke University Press   Durham and London   2016

Library of Congress Cataloging-in-Publication Data
Names: Ong, Aihwa, author.
Title: Fungible life : experiment in the Asian city of life /
Aihwa Ong.
Description: Durham : Duke University Press, 2016. | Includes
bibliographical references and index.
Identifiers: LCCN 2016021420 (print)
LCCN 2016023363 (ebook)
ISBN 9780822362494 (hardcover : alk. paper)
ISBN 9780822362647 (pbk. : alk. paper)
ISBN 9780822373643 (e-book)
Subjects: LCSH: Biopolis (Singapore) | Research parks—
Singapore. | Medical sciences—Research—Asia. |
Biotechnology—Political aspects—Asia. | Biotechnology—
Moral and ethical aspects—Asia.
Classification: LCC HD9999.B443 S555 2016 (print) | LCC
HD9999.B443 (ebook) | DDC 338.4/76151095957—dc23
LC record available at https://lccn.loc.gov/2016021420

Cover art: A sculpture at Biopolis, 2003. AP Photo /
Wong Maye-e.

In memory of my parents
Ong Chin Seng and P'ng Hooi Kean

ENIGMATIC VARIATIONS

Beyond Fortune-Cookie Genetics

In September 2013, an invitation to order a "your 23andMe kit today" arrived at my home in Berkeley. 23andMe is a personal genome service company that was cofounded by Anne Wojcicki (who is related to a founder of Google) in the heart of Silicon Valley. The letter claims that "the service reports on more than 240 health conditions and traits, including carrier status, disease risk and how your DNA may impact your overall health."[1] Furthermore, it added, "You can also learn about your ancestral history." This marketing gimmick underlines that "preventive health information should be accessible to everyone," thus combining a democratizing accessibility with a sunny injunction to self-management.

23andMe celebrates the dream of making DNA technology relevant to personal health, educational benefits, and cultural self-discovery. At UC Berkeley, some administrators were inspired to adopt this user-friendly approach to spark student interest in modern science. In the fall of 2010, the campus initiated a voluntary Bring Your Genes to Cal program. Incoming freshmen were invited to send in their saliva samples to be tested for different kinds of enzyme intolerance.[2] Meanwhile, 23andMe has been promoted in American popular culture for its power and potential to help individuals search for unknown ancestors. A television show on PBS hosted by the Harvard historian Henry Louis Gates Jr. used 23andMe kits to trace the genetic ancestry of famous individuals, stirring widespread interests among African American people seeking to rediscover family lines disrupted by the kinship-shattering cataclysm of slavery. But despite concerns that exposing personal features to the public may lead to social discrimination,[3] personal genetics, packaged and exemplified by 23andMe's merging of consumer empowerment and genomic self-knowledge, is publicly touted as the intertwining of American ingenuity, democracy, and individualism, all mined through individual bloodlines and genomes.

This popular image of genomic science was dismissed as "fortune-cookie genetics" by Dr. Edison Liu, then the lead scientist at Biopolis, Singapore's ecosystem of bioscience institutions. He explained that the growth of personalized genetics companies in the United States has generated the private misuse of genetic information for clues to personal ancestry and health. While 23andMe, for Liu, a U.S. citizen, represents a typically American genomic preoccupation with individualistic conceptions of kinship and descent, he had some reservations. The fact that most people are unable to interpret the data without the intervention of physicians means that the self-knowledge acquired from a cheek swab is not useful from a medical point of view, and indeed it might even encourage individuals to make health decisions without consulting with medical specialists. Indeed, Liu's position was echoed by U.S. doctors and the American Food and Drug Administration (FDA), which disapproved of individuals learning about their own DNA for these reasons. In 2013, the FDA sought to curb the misuse of commercialized, personalized test kits that had led some individuals, on their own, to seek out serious medical procedures such as a radical mastectomy.[4] For Liu, the market packaging of user-friendly DNA is a neoliberal capitalization on individual desires for fortune-telling that only contributes to the fortune of companies and perhaps to the detriment of falsely empowered individual patients.

By invoking 23andMe, Liu seized the opportunity to differentiate an American use of genomics, which seems to project rugged individualism and valorized self-care,[5] from Biopolis, where genomics are managed by scientists for collective health needs. Although the Biopolis hub is closely informed by American scientific administration and practice, as the hub's spokesman, Liu sought to highlight a defiantly Asian difference. As a state-funded project, the Singapore genomics initiative began earlier (2003), intending not to promote personal genetics, but rather to connect genetic data and tissues already stored in hospitals and clinics in Singapore and other sites, especially in China. A community of scientists, not private companies, will supervise the work of linking multiple existing data sources in research institutions and filling in the gaps in genomic knowledge about peoples in Asia.

The Singapore biomedical initiative also challenges the fortune-telling belief that the inheritance story is told exclusively by DNA. Liu explains: "We are in a 'new risk genomics' moment because new research shows that our inheritance is infinitely more mysterious than previously assumed in Mendelian genetics." At the turn of the century, the Human Genome Project was intended to usher in a DNA-focused approach to personalized medicine. Soon after, the focus shifted from a narrow focus on genetics to epigenetics, or the study of gene–environment effects on the performance of genes.

Scientists realize that while the genome evolves slowly through centuries, the epigenome, which turns a gene on or off, can change very quickly, within a few generations. The new science is called post-genomics. Liu prefers the term "new risk genomics," which describes a highly interdisciplinary field that includes genetics, epigenetics, biostatistics, proteomics (protein studies), and metabolomics (the study of cellular metabolites). Liu believes that, as a center for the study of new risk genomics, Biopolis has the potential to generate a tremendous amount of digital information that will revolutionize diagnostic and therapeutic methods. The high ambition of this interdisciplinary ecosystem is architecturally rendered as well in the design of Biopolis itself as a network of interconnected research towers.

Yet, despite Liu's rhetorical dismissal of recreational fortune-cookie genomics, some kind of fortune-telling is involved in genomic science, albeit in the abstract language of DNA and mathematics that still manages to work in "Asian" cultural elements. In the post-genomic landscape that Biopolis configures, and indeed mimetically hails through its architecture, it is precisely the attempt to design and then harness the "experimental future"[6] and its fortunes in Asia that is at stake. This book attempts to illuminate what is cosmopolitan science and what are the variations and differences that become coded in Asian post-genomics.

Biotechnologies today are involved in decoding the secret workings of the genome and recoding it in relation to other systems of codes and information (e.g., ethnicity, disease, nationality, geography). Genetic technologies can be likened to the Enigma machine used during World War II, a device for coding and decoding secret messages.[7] As in the mid-twentieth-century coding industry, the contemporary biomedical enterprise is resolutely multidisciplinary, driven by biological research and bioinformatics. The research milieu is a strange place where mathematicians, biologists, engineers, and other scientists work in tension and in concert across different fields.

The work of unlocking the enigma of life—the double helix of science and passion—now includes research venues in Asia. At Biopolis, DNA databases are coded to "Asian" ethnicities and other elements, thereby redefining what "Asian" means in variations of genes, identity, disease, and space. As a supplement to the American paradigm of the new genomics, researchers in Singapore are amassing and gathering for the first time millions of data points on Asian vulnerabilities and variations, so that other scientists can develop drugs and therapies tailored to the needs of bodies within Asia. I seek to illuminate one of the latest iterations of a century-long migration of scientific and technological knowledges originating in Europe and the United States to Asia, and the situated discovery of new findings within particular biomedical assemblages that transform contemporary science.

Asia, Anthropology, and Science Studies

The path for the study of post–World War II science, technology, and medicine in East Asia was blazed by anthropologists conducting research on Japan, arguably the most scientifically advanced nation in the region. In a pathbreaking study of high-energy physicists in Japan and the United States, Sharon Traweek examined the social and discursive construction of scientific communities.[8] Margaret Lock's award-winning studies of aging and menopause, as well as of organ transplantation, also situated biomedical innovations within a Japan–North American framework.[9] Arthur Kleinman pioneered the cross-cultural study of health practices by contrasting Western and Chinese-style approaches to psychological illness in Taiwan.[10] In a similar cross-cultural vein, Lawrence Cohen explored the medical and cultural construction of senility and cultural anxieties in India and the United States.[11] By taking a comparative approach, these works highlight Asian cultural notions of community, sickness, and bodies that contrast with American scientific understanding. Collectively, such perspectives situate Asia within contrastive cultural contexts for modern sciences.

More recent studies about how scientific and medical knowledges are taken up in diverse regions tend to focus on exploitation and ensuing ethical dilemmas. Brandishing the notion of "biocapital," Kaushik Sunder Rajan framed India as a site that has been exploited by biomedical trials in search of readily available experimental subjects.[12] Other anthropologists have portrayed Asia as a region of coerced and illicit organ harvesting, supplying body parts for transplant procedures, as well as a site of affective labors that serves a burgeoning medical tourism industry.[13] The implications are that besides the "bio-availability" of exploitable populations, cultural and social arrangements in parts of Asia abet in the biocapitalist pursuit of readily available bodies, labors, and "fresh" human organs from the developing world.

Meanwhile, the rapid deployment of specific biotechnologies in Asia requires a shift from contrastive cultural or political economic comparisons, to consider emerging competitive scientific milieus in their own right. The volume *Asian Biotech* casts light on the varied deployment of biotechnologies in Asian sites and on their enmeshment with situated forms of nationalism, biosovereignty, and ethics.[14] The newly influential journal *East Asian Science, Technology and Society* publishes articles that attempt to discover similarities and differences in the production of scientific knowledge in various historically situated but globally enmeshed contexts. Indeed, researchers in the anthropological and science and technology studies (STS) fields are studying emerging science contexts in Asia, which can generate potentially critical insights that richly expand the field beyond its originating Euro-American context.

Framed by the concept of "global assemblage,"[15] this book identifies an emerging context of what may be called Euro-American cosmopolitan science, crystallized in Singapore. First, assemblage concept departs from simplistic cross-cultural and North–South contrasts; it also challenges the STS theory of a universal science that floats beyond local mediations. The emergence of a science milieu in Asia, I argue, is the particular outcome of complex mediations between global technologies and situated forces. Second, if we understand Euro-American cosmopolitan science as regulated science, one should not assume in advance that biomedical science in other places is merely a debased form. Rather, this work will illuminate how, in order to become universal, cosmopolitan science must remediate situated elements so that it can attend to an array of "global" scientific problems. What is "global" and what is "situated" are destabilized in processes of scientific remediation across the planet. In order to be universalizable, cosmopolitan science depends on this constant effort to be particular, to remediate situated elements.

Radical uncertainties, the historian of science Steven Shapin observes, attend much of contemporary science, and "it is the quotidian management of those uncertainties"[16] that is the stuff of my investigation here. My overarching theme is productive uncertainty, in that scientific practices responding to myriad challenges are productive of new forms that in turn create uncertainty.

Different registers of uncertainty are at play in conditioning the experiment at hand: from the calculation of genetic risks for diseases, to uncertainties surrounding the science and the endeavor, to the larger "known unknowns" that science confronts in attempts to secure the immediate future.

Here I take the opportunity to state that, as an anthropologist, my task is to report and interpret scientific practices and ideas in context, without advocating on behalf of actors or experiments under investigation. My approach has consisted less in judging ethical or redemptive claims about specific research objectives than in identifying the particular biomedical assemblages within which ethical problems and conundrums crystallize, which actors seek to resolve. By offering a multifaceted ethnography of bioscience at Biopolis, I aim to illuminate how science projects are complex entanglements of reason and the passions. The branding of a new biomedical center is often surrounded by promotional publicity. As such, media stories and hype are part of the affective work of the trust-making necessary for garnering legitimacy for this kind of state-supported scientific enterprise. Discursive and nondiscursive practices surrounding Biopolis illuminate what might be called a form of scientific "exuberance"[17] as well as the affective uncertainty that perturbs the orderly landscape of science.

At Biopolis, scientific entrepreneurialism as a mode of risk-taking seeks to shape an emerging region for health markets and biosecurity. This ambitious and potentially risky project is inextricably linked to narratives that establish a spectrum of "Asian" differences—in DNA, populations, disease risks, disease forms, geography, research capacities, customized therapies, markets, and collective goals. The remarks of scientists and physicians accord value not only to themselves as experts, but also to the techniques and procedures involved in the acquisition of these truths.[18] My informants often make optimistic projections about the novel value of their discoveries and techniques for "Asian" peoples, the region, even the world. Such narratives and claims are consequential: the regime of truth accepts and makes true the critical potentials of their science.

In addition, science discourses and metrics are strategic when lab findings migrate to the public realm, and science spokesmen must perform in order to continue to draw multibillion-dollar investments from the Singaporean state and from foreign entities. Collectively, promissory claims about the science being produced animate political interest and legitimacy in what citizens may view as an uncertain economic enterprise. Such political justifications have scientists posing the need for Biopolis and the post-genomic research that occurs there in relation to the many diseases and ailments that vex and will vex Asian bodies. To gain further traction, long-standing notions of Asia, now reworked as a genomic, epidemiological, and environmental continuity, come into play. In Singapore, discourses of cultural, ethnic, and geographic differences are less about cultural jingoism than strategic claims to leverage Singapore's potentialities in global genomic science while also making the state investment in biomedical research also a reinvestment in the well-being of a vulnerable and racialized populace.

Race and Ethnicity in Medicine

The United States is a major shaper of cosmopolitan science, but it suffers from the historical convergence between structural racism, medicine, and biology that has had a devastating impact on minority populations. The history of misuse and abuse of racial data in medicine, with actual instances of eugenic and racial violence, is well attested.[19] Owing to this history of race science—one that medical anthropologists have at times participated in—racialized medicine in America is often read as an insidious and virulent science-as-racism.

As many STS scholars of the history of American racial science have argued, race was never about nature or biology in the first place. Race itself was always "interpretive," or a cultural construction, so to speak. Critics have argued that the uses of race were and are always confused about the genetics of populations, the genetics of race, and the genetic and social causes of diseases. Therefore, the reintroduction of race as a biomarker in genomic science has stirred old fears of the biologization of race, its stigmatization, and this reinforcement of social inequalities.[20] In *Backdoor to Eugenics*, Troy Duster explores the troubling social and ethical implications of genetic technologies, including the misuse of genetic theory and information, on minority groups such as African Americans.[21] Especially among those working with populations that have and continue to be drawn into a new constellation of race and medicine in the United States, rightful skepticism continues, despite the fact that the new "ethnoracial" category incorporates the interplay of nature and nurture into medical research.

Indeed, genomic medicine has propelled the transition from race to ethnicity, thus effecting a different kind of interpretation of disease vulnerability, though the race-ethnicity divide is neither finite nor entirely clear. The employment of the ethnic heuristic should perhaps not be considered as a restoration of scientific racism in genomic science, but as a new technique that is intended to be inclusionary in the mobilization of health data. The

National Institutes of Health (NIH) Revitalization Act of 1993, Margaret Lock and Vinh-Kim Nguyen note, promotes the use of race (and gender) as a scientific category in DNA sampling. They are careful to note that "population," "race," and "ancestry" (the preferred term) that variously correspond to U.S. census categories are not considered discrete dichotomous variables but are used as heuristic devices for studying the frequency of specific genetic traits. This represents a gesture on the part of the NIH at navigating the fraught historical and political terrain in which "race" in its molecularized form has often been read as a causal explanation of historical and ongoing structural social inequalities.[22] Duana Fullwiley argues that the "molecularization of race" can be viewed as intended to rectify the systematic exclusion of gendered and raced minorities in American health research.[23] The ethnic heuristic—mobilizing ethnicity in an experiment as an interpretive tool rather than as a claim to some stable and preexisting biological reality—is one way in which researchers attempt to elaborate a bioscientific enterprise that can include questions of human difference without defaulting into the pitfalls of scientific racism and racist genetic determinism.

Ambivalence remains over the use of ethnoracial genomic data because of its unintended effects on racial politics. Even Lock and Nguyen worry that DNA fingerprinting outside the lab may give rise to biomedical practices that unintentionally promote racial stereotypes, affirm ethnoracial differences, or further commoditize racial medicine.[24] At the same time, despite risks of exacerbating racial blaming and oppression, there is a growing consensus that the use of such genetic markers should be dropped.[25] After all, besides their application as a mode of biomedical inclusion, ethnoracial categories may contribute to social healing in that minority groups, through their biomedical racialization, are finally receiving the sophisticated medical attention they have long deserved. Alondra Nelson has argued that commercialized ethnic DNA can be used as building blocks for projects of reconciliation and thus may be viewed as positive elements for the future of American racial politics.[26]

As I will argue in this book, the ethnic heuristic as an inclusionary aspect of DNA fingerprinting is more unambiguously embraced overseas as an advantageous aspect of genomic science that gives texture and robustness to the DNA maps of global populations so far excluded from genomic science.

"The Difference That Makes a Difference"

We are at a moment when there is a growing international division of knowledge and labor as well as a pluralization of the life sciences. Genomic science is a novel experiment in the interplay of biology, race, and the environment, but each national setting uses different concepts of race (historical, cultural, political, and biomedical) in relation to genomic science for different but not mutually exclusive strategies of bolstering national identity, biocapitalism, and/or biosecurity for the future.

Scientists seeking to configure new knowledge systems outside Euro-American milieus generate what Gregory Bateson calls "the difference that makes a difference."[27] Different systems constantly experiment with form where the constant value is not a thing but a contingency. Drawing on ecology and biology, Niklas Luhmann argues that in society's self-referentiality and future elaboration action is communicated through the constant creation of otherness (contingency) in relation to things that already exist. As is often the case, the largest register of difference is the West versus Asia not as stable things but as relationships among shifting contingencies identified in systems making. Differences (race, ethnicity, geography) therefore are not stable but are rather contingent values that systems use to reduce complexity but end up creating more complexity.[28] Throughout this book, "the West" and "Asia" are invoked by researchers, informants, and sometimes by me in order to indicate the registering of such contingent attributes and relationships from vantage points within different systems of knowledge making (biomedical, political, anthropological, etc.).

Difference and differentiation mark novel aspects of any scientific experiment. When American genomic science is used for non-European populations, race, used as a code for groups with distinctive clusters of genetic, epigenetic, and molecular features, is useful for developing customized medicine. In pharmacogenomics, infinitesimal genetic differences can have significant implications for disease susceptibility and therapeutic responses; and racial/ethnic markers have become a useful technology for sampling populations, testing drugs, diagnosing, and customizing therapies. For instance, variability in DNA and in immunology is scientifically significant in reproductive technologies. Charis Thompson argues that "race" in contemporary biomedical research is a heuristic for identifying the intricate interplay of nature and nurture, of genetics and epigenetics.[29] Thus, attention to "racial" biomarkers of gene–environment interactions is very critical in the success of transplant technologies.

But because race outside the lab can refer to a variety of things, the racialization of genomics often takes on political and symbolic overtones, just as it grows out of fraught histories for creating and classifying human difference.

Different national contexts of genomic science disclose various uses and meanings of race.[30] Latin American countries tend to construct "mestizo genomics" because scientists are influenced by notions of race mixture (from social, historical, and political sources) that come to shape research questions and answers.[31] In Mexico, the digital database is racialized as mestizo or mixed race, in opposition to indigeneity and in acknowledgment of interwoven histories and populations who collectively symbolize the nation. Mestizo blood samples are critical for the Mexican biomedical enterprise because they represent a form of "genomic patrimony."[32] It is interesting that genomic science in Latin America seems to be primarily concerned about constructing unified, while mixed, national races in their databases. By contrast, in Asian biomedical sites, ethnicity as "the difference that makes a difference" is deployed as an astute strategy to enhance the scope and power of genomic knowledge thus generated.

Enduring European colonial legacies in Southeast and East Asia are constructions of plural society, of coexisting races (essentialized) closely tied to language and religion. Different authoritarian political orders are based on multinationalism (China) or on multiracialism (Singapore), and the major axis of difference is between majority and minority nations/races/populations. Although there is political emphasis on protecting the group rights of minority nations/races, the majority nation/race is variously privileged and enjoys political dominance. In Singapore, electoral democracy is tempered by a communitarian ethos that extols social obligations and the importance of the common good, thus emphasizing collective over individual autonomy and rights. An official order of so-called CIMO (Chinese, Indian, Malay-Muslim, Others) multiracialism aims to balance the claims of different races in the nation. At the same time, hate-speech statutes discourage talk about race and religion, and there is a healthy public defense against disparaging the cultural practices of any "race." In this model of administrative homogenization of identities, "ascribed" race minorities are very different from "voluntary" self-inscribed minorities in liberal multiculturalism.[33]

Nevertheless, in reaction against the state's insistence on "racializing" everyone, media, academic, and "scientific" discourses increasingly use "ethnicity." Researchers in Singapore shift from the official category of race (traced through patrilineal descent) to American uses of ethnicity (based on self-identification in medical records) in their effort to model ethnic biomedical collectivities. Fortuitously, they recognize that ethnic-differentiated medical science makes their databases more performative and mobile across multiple sites. For instance, ethnic Chinese biomedical collectivities can come to represent huge numbers of people in the world who may self-identify as Chinese. Critically as well, English—the language of science and ethnicity as normalized by international social science—is utilized to strategic advantage by Singaporean health researchers. The ethnic heuristic helps to circulate their findings, claims, and applications to places where English denotes like-ethnicities are found.

Therefore, genomic science in Singapore does not reify colonial-era notions of biological race, nor does it uphold a single national race in the genomic lab. In addition, the assumed stigmatizing effects of ethnoracial medical data in the United States do not apply in Asia. People tend to have a robust sense of their (variously constructed) racial/ethnic identities viewed through the lens not of past victimization but of ancient roots and historic achievements. Genetic technology is new, and people welcome Asia-oriented research that targets their ethnoracial group for therapeutic research. Few express fear or ambivalence about ethnic specifications in biomedical sciences, which in any case are but tools to help clinicians develop the personalized genetic data one can get on a chip and soon on the iPhone. Ethnic-differentiated tools are part of being techno-savvy medical consumers.

By adopting the ethnic heuristic, Singapore can leverage an ethnic-rich genetic database and brand itself as a biomedical center for a broader Asia. Multiethnic DNA is less about investing in national unity (as in the Mexican case) than a pragmatic strategy to produce a statistical infrastructure for demographic and geographical reach. It is this convergence of the use of ethnic heuristics in cosmopolitan science and the existence and malleability of official racial classifications in Singapore and Asia through which this infrastructure emerges. Racial categories for population administration provide a convenient and salutary statistical framework for the biomedical sciences. Biopolis's American-style biomedical research is thus resolutely global in its ambition; and the ethnic heuristic, detached from specific national moorings, facilitates a transnational inclusiveness because majority populations (Chinese, Malays, and Indians) in the region who were previously excluded from "universal" biomedical research can now be brought under the molecular gaze. In recognition of this universalizable power of the ethnic heuristic, the NIH selected Singapore's "trans-ethnic" DNA project to develop statistical research on the DNA of "non-European" populations.[34] In a sense, American scientists furnished the ancestry/ethnic heuristic, as Lock and Nguyen have argued, and their Singaporean counterparts apply it to majority (not minority) populations in Asia.

This book is an experiment in what I call an anthropology of the future. How can anthropology—the study of the diverse ways of being human—be made relevant in the twenty-first century? Whereas anthropologists have long assumed that "culture" has always had a monopoly in defining the human, Stephen J. Collier and I maintain that science and technology actively mediate cultural notions, thereby proliferating novel ideas of the human, living, and life itself. The task of anthropology therefore is to investigate how contemporary science participates in and transforms preexisting cultural ideas about the anthropos in multiple registers today.[35] In an age of hopes for science and technology, ethnographies are critical for illuminating how cultural, philosophical and political differences translate and shape experimental systems and milieus.[36] Following a visit to China, Nikolas Rose has observed that the racializing trend of pharmaceuticals in Asia should not be dismissed as due to simply cultural differences. Instead of a reflexive critical suspicion, he cautions, we might seek answers in "new relations of genomics, identity, biosociality, and bioeconomics."[37]

In the chapters that follow, my study of Biopolis in Singapore, with a glance at BGI Genomics in China, goes beyond cross-cultural and cross-disciplinary translations to interrogate how science itself becomes transmuted in the process of designing anticipatory futures. This book is an ethnographic study of Biopolis, Singapore's City of Life, a global milieu that seeks not only to incubate a new life science in and of Asia, but also to mobilize new political and ethical horizons for managing uncertainties in a uniquely connected and vulnerable region. Even as therapies are becoming more and more individualized for the wealthy, as in the sequencing of Steve Jobs's genome in order to treat his pancreatic cancer, pharmaceutical innovations continue to demand the capture of huge swaths of new data. But whereas biomedical science is amazing in promising to unlock the codes of life, our diverse and shared fortune as anthropos is not so easily predictable or prepared for.

The new biology evolving in Singapore and elsewhere is an interdisciplinary field, bringing together the diverse expertise of biostatisticians and classically trained biologists, engineers, and doctors who often do not see eye-to-eye but do depend on the same sources of state or overseas funding. Different techniques are fashioned from dry labs and wet labs: that is, sites for the analysis of computer-generated data and classic bench-top experiments with biological materials. My investigation focuses on some research programs integrated with clinical and academic research communities, including genetics, oncology, stem cell research, and tropical diseases. I explore the biomedical assemblage from the inside to illuminate how the work of science is infused with intensities, optimism, and anxiety.

As part of its quest to be a global biomedical hub, Singapore shifted from a British medical tradition focused on high-quality patient care to an American style of training physician-researchers engaged in innovative evidence-based practices. In 2003, Biopolis was established by the Agency for Science, Technology and Research (A*STAR). Biopolis comprises a cluster of public research institutions and corporate labs involved in many areas of biomedical science activities. Outside the Biopolis precincts, there are many international medical programs, including the Duke-NUS Graduate Medical School and the Johns Hopkins Cancer Center as well as major teaching hospitals and global drug laboratories. Biopolis is then itself less a singular site and more a network of institutions stretched across the island and beyond. With the term, "Biopolis complex or ecosystem," I refer to this extended network of universities, hospitals, clinics, research institutions, and pharmaceutical companies in Singapore and overseas.

Singapore has gathered an international community of life experts (biostatisticians, geneticists, stem cell experts, neuroscientists, bioethicists), the so-called new specialists of the soma,[38] to meet such challenges. The bioscience research community draws from the public and private sectors, composed of more than two thousand scientists. Foreign and local-born researchers have been trained at leading world institutions such as Cambridge University, University of Edinburgh, Harvard University, MIT, Johns Hopkins, and many more in Europe and Australia, as well as Singapore's own world-class universities. Science luminaries supervise labs, unfairly dubbed "research factories," where hundreds of PhDs recruited from top-ranking universities in China, India, and Singapore work in some obscurity. Despite their busy schedules of work and travel, all scientists whom I contacted were responsive to requests for interviews. Biopolis has many corporate labs, but scientists there were unavailable for interviews because of concerns about intellectual property issues.

This book draws on research conducted between 2004 and 2013 during multiple summer visits to Singapore. In all, I interviewed a few officials and scores of researchers in fields such as population genetics, medical genetics, oncology, bioethics, infectious diseases, and stem cell research in the extended Biopolis complex. My investigation focuses on research practices rather than on therapeutic activities, and my informants tend to be scientists (principal investigators) who often are clinician-scientists. Most of my interview data were collected in the spring of 2010, when I was a research fellow at the Asia Research Institute of the National University of Singapore. Some scientists were interviewed later at UC Berkeley and the UC San Francisco Medical School in California, and BGI Genomics, China.

Besides hour-long interviews (and repeat visits in many cases) at the offices of science institutes, I attended the many international conferences and lectures at Biopolis and the Duke-NUS Graduate Medical School. I also visited major teaching hospitals and clinics throughout the island, and I generally imbibed the biomedical culture brewing in Singapore. I hung around different medical campuses and ate in cafeterias serving international cuisine. This fieldwork, driven in part by my capacity to connect with individual researchers, offers captivating ethnographic and philosophical moments that highlight the invisible work, as well as the uncertainty, going on in some of the labs.

I am grateful to all respondents, from principal investigators to lab workers, from American scientists to mainland Chinese technicians, for their desire to explain to a nonspecialist what it is they are doing. I was generally impressed by their ardent interests, strong dedication, and professed optimism for the future. The identities of informants are disguised except where otherwise indicated. Scientists with public roles and well-known reputations—such as Edison Liu, director of the Singapore Genome Institute (2003–2010), and Henry Yang Huangming, a founder of BGI Genomics, among others—retain their own names. I appreciate the time and effort they took to engage someone who is concerned about the anthropos in other guises.

Not all scientists I encountered participated in the project of ethnic-stratified medicine, and many projects at Biopolis do not mark their data or claims in ethnic terms. But as one among other Asia-born researchers, my presence may have stimulated a degree of candidness seldom encountered by other anthropologists. In Singapore, cultural discourses suggest an overlap between race and ethnicity, and that will be evident in quotes scattered through this book. At the same time, most researchers frequently invoked "Asia" and/or "Asian" to highlight some dimension or element—in genetic variants, beliefs, values, way of life, and geography—that is a necessary and significant part of their work in forming this globalized biomedical milieu.

ACKNOWLEDGMENTS

I thank Dr. Gregory Clancey, the leader of the STS research cluster at the Asia Research Institute (ARI), National University of Singapore, for welcoming me as a senior research fellow in 2010. Dr. Chua Beng Huat, the chair of sociology/anthropology, and Dr. Lily Kong, at that time the director of ARI, were gracious in inviting me to join in the academic and social worlds of Singapore. I certainly developed an appreciation for the ambitious effort involved in navigating a shift from British to American medical practice, as the Biopolis venture is perhaps a not-so-surprising science strategy for reimagining the present-future of this metropolis, and perhaps the region.

Back in Berkeley, I thank the Institute of East Asian Studies and the Center for Chinese Studies for funding multiple phases of research and writing. I am especially grateful to Jerry Zee and Gabriel Coren, who read final, final drafts of the manuscript. Their helpful questions and suggestions, as well as moral support, kept me going. Early versions of some chapters were read by Andrew Hao and Limor Darash-Samarian. An earlier version of chapter 1 was published as "Why Singapore Trumps Iceland: Gathering Genes in the Wild" in *Journal of Cultural Economy* 8, no. 3 (2015): 1–17.

I appreciate the two anonymous reviewers whose comments helped clarify the structure of the book. Thanks to Ken Wissoker, press editor extraordinaire, for his continuing interest and patience. As always, I received the unstinting support of Robert R. Ng.

The practical circumstances in which I first started exploring the rise of Biopolis were generated by my visits to my mother in Singapore. Traditional Chinese acupressure, more than modern medicine, offered comfort in her final years. This work is dedicated to her memory, and that of my father.

INVENTING A CITY OF LIFE

Others collect butterflies; we collect scientists.
—SINGAPOREAN OFFICIAL (2005)

Biopolis is a life-sciences hub in Singapore that is at once embedded in the Asian tropics and densely connected to biomedical science sites around the world. Conceived and implemented by a Singapore government body called the Agency for Science, Technology and Research (A*STAR), Biopolis is the heart of a new bioeconomy built to remake the near future. The galactic imagery of A*STAR is reiterated in the words of a Singapore leader who boasted that this port city must be like a Renaissance city-state (i.e., it must become a crucible of creativity that thrives by welcoming talented people from far and wide).[1] Biopolis, which is central to Singapore's reinvention as a knowledge economy, was introduced to the world with extravagant flourishes and fanfare.

In the first two decades, the Singapore state poured billions of U.S. dollars into the biomedical center at the One North campus. Biopolis began in 2003 with an initial cluster of nine interlinked towers—there are now thirteen—dedicated to bioscience activities conducted mainly through public research institutes. The image and tone of the place were established by the international architect Zaha Hadid, who helped design a stunning parkland for scientists to work in. The key research towers are named after Greek mythological figures—Helios, Chromos, Centros, Nanos, Matrix, Genome, and

FIG I.1 A "sky bridge" hovers above a tropical garden at the Biopolis complex.

Proteus—signifying the high ambition and international symbolism of the projects.[2] These public institutes, increasingly juxtaposed with corporate labs, are engaged in cutting-edge research in genomics, stem cells, oncology, neuroscience, nanotechnology, and biologics as well as tropical diseases. The towers are linked by sky bridges, an architectural rendition of connectivity, to reflect the resolutely international and interdisciplinary orientation of the initiative (see figure I.1). Visiting British scientists, impressed by seemingly unlimited funding, top-notched equipment, and spectacular facilities at a time when funding for science has become less certain elsewhere, have dubbed Biopolis a "science nirvana."[3]

In the early years, the hothouse atmosphere was underlined by claims that Biopolis was no butterfly-collecting expedition, but instead a project to collect scientists. Alan Coleman—the Scottish scientist who famously cloned the sheep Dolly and now leads a program in stem cell research—was the chief representation of the kind of "world-class" expert that Biopolis aimed to "collect," who then acts as a principal investigator (PI) for different institutes and programs. The scientists oversee laboratories filled with hundreds of PhDs from China, India, and other parts of Asia who have been offered multiyear A*STAR fellowships. Recruited by headhunting programs, these lab researchers are well paid compared to those at other research centers in Asia and, once in Singapore, they are encouraged to take up citizenship. Talented Singaporean students are sent for overseas training in science and engineering, but they are expected to return to work at Biopolis. While the goal for the future is to have the biomedical hub be mostly homegrown, the community of scientists is currently international, and their work is to advance cosmopolitan science in the Asian tropics.

Biopolis is dedicated to the bright promise of developing personalized medicine in Singapore. It is part of an ambitious quest to code variations in DNA, ethnicity, disease, and location among Asian populations for the discovery of novel genomic information. Like variations in a piece of music, scientists' refrains often invoke the difference of ethnic- or Asian-stratified medical data that brands Singapore as the prime milieu in which global pharmaceutical innovations can be made in Asia and for Asian populations.

While the ethnicization of genomic data for customized medicine in the tropics remains a heuristic, utilitarian way of discovering and investing in genomic citizenship or Asian genes, at the same time, the data becomes "Asian" as does the modality of research and of life science. That is, Biopolis operates not only as the center of a new research ecology but also as a key site in the staking of a new and self-consciously Asian way of doing science. This begs the question as to why the science becomes veritably Asian while still being international and cosmopolitan in its design and practice. At stake in this biomedical assemblage is the crystallization of conditions for the cosmopolitanization of a science that now refers to Asian bodies/histories/migrations/diseases.

A City of Life

A self-description of Singapore is that it is a tiny, resource-poor island nation that is compelled to constantly self-invent. Since its independence from British colonial rule in 1959, the city-state has struggled to survive. The 1960s were a fraught decade, characterized by an ill-fated union with Malaysia that ended in 1965. The island nation also had to cope with a "konfrontasi" (Malay-Indonesian) policy from its giant neighbor, Indonesia, which reviled independent Singapore as a running dog of Western imperialism. Under the extraordinary leadership of the first prime minister, Lee Kuan Yew, the next few decades saw the stunning rise of Singapore as an "Asian tiger" nation, leaping from being a manufacturing center to a global port and financial hub. By the turn of the century, Singapore's GDP per capita of over US$55,000 exceeded that of the United States.[4] Orville Schell, a scholar of modern Asia, notes wryly that Singapore's experiment with modernity made "autocracy respectable" by leavening it with meritocracy.[5]

The modern history of a tiny, resource-poor island struggling in a hostile ocean has engendered an ethos of *kiasu* (Hokkien Chinese),[6] or "fear of losing out," that pervades public policy and everyday activities alike. The Singaporean version of meritocracy, which derived in part from the Confucian valorization of education and from the modernist focus on progress through expertise, has fueled a *kiasu* as an effect of fierce competitiveness in order to avoid "losing" in individual as well as government ventures. The nation's leading sociologist, Chua Beng Huat, argues that "fear" of failing to win haunts the success that has become the Singapore identity and brand.[7] Not surprisingly, an undercurrent of anxiety suffuses state entrepreneurial projects such as Biopolis. Especially since the SARS (severe acute respiratory syndrome) outbreak in much of Southeast Asia at the turn of the century, the affective effect of *kiasu* has taken on new urgency, driving new senses in the necessity of not only sound, state-led planning but also the need to be vigilant, if only to avoid anticipated disasters, including those of a biological nature. The turn to the life sciences has taken the form of Singapore being a beachhead for American cosmopolitan science, while the influx of U.S. science institutions seems to register an American anxiety about sustaining influence in the Asia-Pacific region as well.

In recent years, with an eye to the rise of China and India, the Singaporean state has shifted away from manufacturing to focus on high value-added industries. In economics, "value-added" refers to the increase in value of a product, exclusive of initial costs, at each step of its production. Knowledge and informational technologies, by enhancing manufacturing, marketing, processing, and services, are ways to add value to a product. With some of the highest student achievements in math scores in the world, Singapore has rebranded itself as an "intelligent island." The Economic Development Board began to quickly step up investment in research and development generally, especially in projects that promised to have a "high multiplying effect" in stimulating the growth of a knowledge economy.

The quest for new sources of value in the midst of anxiety over emergent viral and biological threats also prompted a refashioning of citizens as "brain workers" who are urged to reject lucrative jobs in finance for occupations that take care of "sick bodies."[8] The shift, from treating the population as an ever-productive labor force to a pool of bodies that will be the source of diseases and of novel medicines, is dramatic. With its efficient system of public health financing, and the recent computation of multiracial medical data, the Biomedical Research Council (an arm of the A*STAR galaxy) sought to reposition Singapore as a biomedical research hub and a health destination. In 2003, the SARS epidemic unleashed fears of not being prepared to deal with health epidemics looming for tropical Singapore and, as a regional transport hub, its far-flung environs. SARS threatened to derail the economies of Asian nations and grew into a pandemic that menaced the rest of the world. Because SARS was an "Asian" disease and lives lost were initially mainly in Asia, the perception of researchers and physicians as virtuous public servants is closely tied to regional and national identities. "SARS," a leading Singaporean epidemiologist confided, "helped the government to convey the message that nothing can be taken for granted." In the aftermath of SARS, a new vigilance about potential contagion threats shifted Biopolis from a center narrowly focused on shaping a bioeconomy to being on the frontlines in the fight against infectious diseases in the region.

This effort has been closely tied to new articulations of biomedical sciences, based in Singapore and, ultimately, science itself. While building a research platform for novel problematization of and intervention into "Asian" bodies, citizenship, and well-being, Biopolis has also staked its ambitions in cosmopolitan science. After all, from its beginnings, the initiative was advised by a group of well-respected experts from the United Kingdom and the United States. Among them was Dr. Sydney Brenner, a Nobel Laureate and pioneering molecular biologist who joined the Singapore National Science Council as a consultant on Biopolis.[9] As one of the pioneers of genomic science, Brenner made insightful and ethical interventions into our hubris regarding what we know with the knowledge we make in the life sciences. By having ethically minded star scientists on board, the Biopolis initiative aimed to demonstrate a dedication to science and a desire to learn and self-cultivate science as an enterprise, in the sense of to invent and create, beyond the crass materialism or bald global ambitions through which Asian sciences are often dismissed. Therefore, despite the media hoopla attending its early years, the Biopolis endeavor cannot be reduced to a purely entrepreneurial project with a still murky future. Some fitting questions posed by Biopolis may be how is scientific knowledge governed at a global scale, and what are the implications of a novel and distinctive "Asian" model of knowledge production for understandings of life?

This book illuminates a charged Sputnik moment in contemporary Asian bioscience[10] when scientists at multiple sites are experimenting with different visions of the future. Anthropologists often view biomedical innovations as contributing to the exploitative dynamics of biocapitalism,[11] or at least driving predatory practices of "bioprospecting,"[12] trends that variously intensify inequalities between rich and poor countries. At the same time, we cannot ignore how shifts in the biomedical industry beyond blockbuster drugs have opened up new opportunities for emerging countries to gain some control over their biological resources and secure the well-being of their peoples. For instance, in 2008, when Indonesia famously and controversially refused to share H5N1 viral specimens with the World Health Organization, observers considered this refusal an economic ploy to seek payments. Yet the outcome of Indonesia's negotiation with drug corporations that used the diseased samples was to provide vaccines at lower costs to donor countries, thus benefiting their citizens.[13] This is a critical example of how a big pharma–dominated notion of the "global good" is mediated by the interests of emerging nations.

The worldwide dissemination of biomedical tools and drugs has the capacity to generate a range of potential values, and only some can be construed as potentially "economic" in a strict sense. Indeed, this excess of value over the narrow constraints of classical economics is a key observation of the discipline of anthropology in general. Things being traded cannot be reduced to sheer commodities, but continue to bear the aura of social relationships and are thus animated by complex meanings and obligations.[14] Even in the era of big pharma, we may still hesitate to make a value judgment in advance and instead explore what clusters of values are in formation with the circulation of drugs and biotechnologies, what valuations are at stake, and how small countries can negotiate and constrain the power of global corporations. This more situated approach allows the anthropologist to evaluate the worth of what post-genomics science can enact, what vital investments it can make, and what hopes and dangers it can instigate for the collective good in emerging regions of the world.

It is also clear from earlier sales pitches that Biopolis positions itself as a strategic hub leveraging Asia as the world's next big drug market. Nevertheless, the creation of novel knowledge in a biomedical frontier begs the question of how the interrelation of biotechnologies, capitalism, and politics can also be generative of alternate goals. That is, if capitalism, geopolitical inequalities, and knowledge co-constitute the space in which Asian biotech aspirations operate, do they also open possibilities for other hopes and goals not overdetermined in advance? Biopolis, as I will illuminate, is not just an ecology to generate a particularly active form of scientific life,[15] but also a research milieu oriented to its tropical setting, peoples and other living forms, and closely tied to strategies for repositioning and remaking Singapore, and the Asia it represents, into a major scientific and medical player globally. Rather than invoking a new epoch in the rapacious and auto-elaborative agency of capital, I explore Biopolis as a contingent juncture of various processes and elements, of which capital is one, by attending to how practices of calculating and managing uncertainties produce enigmatic analogs of life in and of "Asia."

Situated Cosmopolitan Science

Scholars of science have noted that specific modes of scientific cultures and objects are shaped within various political and research environments. Lily E. Kay analyzes the interactive elements that gave birth to molecular biology in the United States,[16] while Sheila Jasanoff compares the varying impact of democratic citizenship, public culture, and nation-building endeavors in differently shaping science policies in North Atlantic nations. She notes that variable political cultures condition distinct research apparatuses, which might be understood as part of "projects of reimagining nationhood at a critical juncture in world history."[17] In a more ambiguous formulation, Hans-Jörg Rheinberger observes that the history of methods, objects, and key sites of experimentation in genetic and molecular science suggests multiple ways in which such experiments are crystallized. He maintains that "assemblages—historical conjunctures—set the conditions for the emergence of epistemic novelty."[18] This exploration of the biomedical scientific enterprise in Singapore provides a significantly different picture than studies of bioscience cultures in Euro-American environments.[19]

How to bring together—in a particular configuration of cosmopolitan science—the epistemic novelty of post-genomic science on the one hand and the situated political conditions on the other hand? How do situated political and ethical re-imaginings work to impact the *novelty* of epistemic novelty, or shape these novelties, as it were? Stephen J. Collier and I have offered the idea of the "global assemblage" as a useful lens for identifying the complex interactions of global knowledge and technologies on the one hand and situated contexts of politics and ethics on the other. Specific articulations of global and particular forms, we maintain, crystallize situated circumstances for generating novel concepts, objects, and tools for solving problems of life and living. As an alternate to conventional units of analysis such as the nation-state, empirical assemblages of technologies, institutions, and practices give a frame to emerging situations of problem solving.[20]

The Biopolis complex is formed at the nexus of cosmopolitan science and Singaporean authoritarian politics and collectivist ethos, raising questions about how global and situated elements interact with one another and what effects their adjacency elicits by defining what counts or matters in this form of cosmopolitanism, or what makes it such. The interplay of cosmopolitan life sciences and political entrepreneurialism in Singapore engenders situated problematics of risk in this emerging Asian bioscience. There is the epistemic novelty of a semiotic landscape of biopolitical governance that insists on "Asian" differences for understandings of life. At the same time, there is the development of global scientific capacities to deal with the risk of disease emergence in Southeast Asia that threatens the world. I will set out how these disparate but interlinked strategies for managing health risks complicate the meaning and challenge of science entrepreneurship in Singapore.

Scientific Entrepreneurialism in Asia

Research scientists are artists who push the
boundaries of convention. . . . They are risk-takers
seeking to develop biosecurity in a world of
information flows and fungibility.
—EDISON LIU, director, Genome Institute of
　Singapore (2003–2010)

A model of a long spiral of DNA stands in the lobby of a gleaming Biopolis building, seeming to reach for the heavens. The double helix is both the glistening substance of and symbol for a genomic future. Biopolis's imposing architecture consists of a group of state-of-the-art research institutes housed within interconnected towers. They nestle in gardens designed to suggest a mix of tropical jungle and high-tech nursery, the image of a science designed to intervene in tropical life, scientific sociality, and innovations. Situated on a knoll, Biopolis is part of a larger digital information complex called One North (indicating its latitude north of the equator), Asia's latest venture into the brave new world of life sciences.

Biopolis was shaped during its first decade (2003–2010) by an energetic Chinese American oncologist, Dr. Edison Liu. Liu was the first director of the Genome Institute of Singapore as well as Singapore's chief science spokesman. During his tenure at Biopolis, Liu became the first Asia-based leader of the Human Genome Organization (HUGO).[21] From the start, Liu was a key thinker who envisioned and directed genomic research in Singapore and many sites in Asia. In the quote above, Liu signaled his view of scientific entrepreneurialism, emphasizing the role of scientists as risk-takers who must "push the limits of conventions" in a world of competitive information and value flows.

Genomic information, he argued, must be made fungible across spheres of knowledge, market, and security. Much more was at stake than just calculating bioprofits; governing apparatuses become fine-tuned as problems of biosecurity accumulate in diverse zones. The larger implication in a world of competitive flows is that scientists in the Biopolis ecosystem must create a distinctive space of intervention: one that is differentiated from and can tactically differentiate other contexts of biomedical science. The entrepreneurial goal of Biopolis, as expressed by Liu, is to shape a field of science research in Southeast Asia and beyond that can be the basis of defense against biopiracy so that DNA from Asia is not reduced to a "cash cow" for big pharma.

In the 1980s, a new form of scientific entrepreneurship emerged in the United States when American industries became alarmed by the perceived economic and technological competition represented by Japan and Asian "tiger" economies. The U.S. response was to encourage collaborations between research universities and major industries, and public and private institutions were encouraged to shape the growth of high-tech and biotech regions, starting in California.[22] Steven Shapin argues that entrepreneurial science cannot be disassociated from the charismatic authority of the figure of the lead scientist best exemplified by J. Craig Venter, a maverick scientist who was a key player in sequencing the human genome. Venter went on to found a private company initially called Celera Genomics that has since spun off a number of entities under various names. But as bioscience research went global, scientific entrepreneurialism has come to mean something different in a biomedical frontier such as Biopolis, and the charismatic leader has to be a very different personality type than Venter.

If Liu can be taken at his word, what is the moral lesson for scientists in Singapore that requires them to be risk-taking beyond the lab? The research model is not to be Venter-esque but to constrain the ways biomedicine has become corporatized. Asian researchers are operating in a region where big pharmaceutical companies may seek to colonize bioresources and abstract commercial values to make profits. As state employees, scientists in Singapore wish to take the lead in corralling scientific objects and findings about a variety of life forms in the region before they fall into the hands of drug companies. As we shall see, India and China are very concerned as well about protecting their natural resources and controlling the uses of data and values derived therein.[23]

In emerging nations where science has always been viewed as a tool of emancipation and thus bound up with the fortunes of the nation and

citizens, the Weberian notion of virtue in science as a calling needs quali-
fication and contextualization.[24] In Singapore, most of the scientists employed within the Biopolis ecosystem are public servants. While quite a few
may love science, all are answerable to taxpayers, and virtue in science is
expressed as a public service that cannot be labeled a simple vocation or
as its opposite, sheer careerism. Rather, biomedical entrepreneurialism in
this context is a developmental necessity to manage the collective interests
of peoples in a region on the verge of being invaded by big pharma. The
synergy between fear of failure and collective fate is expressed in contrary
affects of anxiety and hope. The science experiment becomes inseparable
from the performance of civic virtue and its investment in biological futures
of the emerging world.

The interplay of hazards and hope, as well as uncertainty of outcomes,
fuels "promissory" claims[25] about such an expensive biomedical initiative.
It becomes the moral role of science leaders to stir up public enthusiasm
and legitimacy for a state-funded, science investment in the collective good.
Thus, bioscience entrepreneurship in Singapore is not best understood as
a corporate strategy to shape speculative drug markets in Europe and the
United States by opening the Asian arena. In Liu's discourse, the affective
resonance is about being entrepreneurial with "Asian" differences—from
DNA to ethnicity to disease to location—that promise to make a difference
in personalized medicine and that resonate with citizens as cultural subjects,
taxpayers, and patients. The Biopolis project must be shown to be bullish
about Asian needs, not just be a commercial outpost of American biosciences. Singapore's scientific entrepreneurialism is in a larger sense about
claiming ownership of a novel science of peoples in Asia, but it does not
mean thereby that it lacks substance. A decade on, the biomedical science
initiative is thriving.[26]

After SARS, Biopolis was recast as a center for biodefense, and "scientific
entrepreneurialism" came to include this public responsibility to prepare for
impending catastrophic events. The Duke-NUS Graduate Medical School was
established to help train researchers to develop expertise on the endemic infectious diseases plaguing Southeast Asia. Biopolis and these other institutions
together play a dual role as biocapitalist and biosentinel, ever alert to managing uncertainties in the market and in nature that threaten the nation and surrounding region. Southeast Asia and its surroundings are an emerging region
of the world, as well as a biologically rich ground zero for the rise of deadly infec-

tious diseases. Biomedical "Asia" is variously problematized as a multitude of risks to be managed through the discovery and oversight of findings about DNA, human and nonhuman life forms, and diseases and biothreats that are considered specific to the region.

Physician-researchers in Biopolis, then, need to demonstrate interest in the public good even as they become entangled with private interests of corporations and professional self-interest. Certainly, given their international training, scientists whom I interviewed in Asia were not immune to the American model of the entrepreneurial scientist who celebrates individual ingenuity. Nevertheless, the talk about risk-taking in Asian bioscience is a reminder that there are larger stakes in "scientific entrepreneurialism" than the quest for professional fame or corporate profits.

Scientific entrepreneurialism translated to Singapore does not describe the kind of bold risk-taking innovations by Venter that rock the new genomics. Rather, clinician-researchers in Singapore, and at BGI Genomics in China, claim that they focus on "practical things and wish to avoid controversial projects" such as creating artificial life. As is the case in many postcolonial countries, modern science is considered part of nation-building efforts, and the state tends to be the organizer of science training and research activities, and scientists tend to be public servants. Science expertise tends to be found in universities, teaching hospitals, and research institutes, the core institutions that institutionalize civic virtue among practitioners. At Biopolis, entrepreneurialism includes the moral expectation of Asian scientists, as leading public servants, to protect citizens' interests at a time when prowling drug companies are both a threat and an opportunity. Whereas Shapin argues that the uncertainties of much contemporary science in the United States have made "personal virtues" *more* central to its practice than ever before,[27] for emerging nations, public virtue remains salient in state-driven forms of scientific entrepreneurialism even in the midst of tempting opportunities offered by pharmaceutical companies. Virtue in the sense of serving collective interests would include responsibilities to produce knowledge on and take charge of national bioresources and patrimonies. As Nancy Chen and I have argued, scientists in emerging Asia are expected to defend their nation's biosovereignty in the face of challenges posed by global drug companies.[28] The public-private partnerships in scientific life are found everywhere, but perhaps in some Asian nations, public interests are necessary to generate affects such as trust and legitimacy.

Pluripotency and Fungibility

Uncertainty, Michel Foucault notes, is itself a form of power, a biopolitics that uncovers the enigma of life and materializes it into a manageable kind of present-future.[29] In Jane Guyer's words, "the near future," therefore, is that space that falls within the horizon of calculability.[30] By falling between the immediacy of the present and the loftiness of distant utopian futures, this temporal scale remains provisionally actionable and within the realm of pragmatic calculation. Genomic science, I argue, participates in this shaping of the present-future, by governing as it were through uncertainty, by calculating health normalities and risks as well as anticipating biothreats that may disrupt the near future.[31]

Furthermore, when we take a situated approach to scientific configurations, we discover glimmers of bioscience reasoning that go beyond what Nikolas Rose calls technologies of optimization focused on managing the arts of a healthy lifestyle. Given the stakes of the life sciences in Asia, "the politics of life itself" is premised on so much more than "what it is to be biological."[32] Biological sciences in emerging sites are perforce oriented less toward self-optimization than technologies for managing uncertainties that threaten in ways large and small the collective interests of life and living in the region. In Singapore, this broader concern with biosecurity may be said to be framed within an emerging form of biopolitical governance and its understanding by the government and civil servants as an ethic of collective care. In the overlapping geopolitical and sociopolitical interests that ride on bioscience in Asia, I argue, researchers are driven by larger goals in an experiment of "making more of life" that goes beyond enhancing the vital future of individual consumers.

Given their location, scientists in the Biopolis ecosystem seek to discover a range of life values from the biodiversity that surrounds them. I invoke the term "pluripotency" to describe the movement from the actual to the virtual, from singularity to multiplicity. In the new induced pluripotent stem (iPS) cell technology, the perturbation of adult cells causes them to revert to earlier embryo-like stem cells that are capable of growing into multiple types. Likewise, I track how scientific practices at Biopolis shift from the undifferentiated to multiple differentiated realizations that can be understood as a deterritorialization of the life sciences through which radical possibilities are unleashed.

I argue that in Singapore a pluripotent reasoning reinvests the collective "Asian body" as a distinctive kind of medical object—not the "universal" raceless body of white or unraced medicine, but as variations of situated, ethnic, and sick bodies. By coding and valorizing genomic and ethnic variability, Biopolis scientists are better able to be competitive in the international arena of bioscience research and pharmaceutical investments while also continually re-emphasizing a reinvestment in the diverse "races" that compose Singapore and Asia at large.

It is the historically contingent composition and racialization of Singapore's citizenry, created through population flows in the British colonial adventure in Malaya (and Dutch incursions into the East Indies) and enumerated through administrative schemes that know the population as a mosaic of official races, that has become a demographic and data infrastructure for building relations with both the medical establishments of other Asian countries and consistency with ethnoracial categories in cosmopolitan science.

Building regionally and ethically varied databases provides opportunities for Singaporean scientists to collaborate across the politically fragmented landscape of Southeast and East Asia. Cross-border science alliances can yield more ethnic-differentiated data and samples, thus creating a foundation for a potential Asian DNA databank. Furthermore, trans-Asian cooperation in science training promotes the beginnings of regional preparations for dealing with epidemics and other anticipated biothreats. The life sciences, in Singapore and China, produce the beginnings of regional collaborations that may come to define the orientation of cosmopolitan science, shaping what Brian Buchanan calls an emerging form of "geo-biosociality."[33]

The coding and alignment of variations are practices that make genomic science pluripotent and fungible. Pluripotency and fungibility in Singaporean genomics operate through the reassembling of existing forms of racialization and racial accounting in the nation's official classification of its citizenry. It builds on the ongoing use of ethnic heuristics inherited from British colonial racial typologies. But while these categories come out of a history specific to British colonialism, they have been increasingly leveraged to position Singapore as a demographic kaleidoscope of the populations of East, Southeast, and South Asia at large. In other words, without recourse to postcolonial theory, or claims about the continuity and sameness of political processes of racialization, these inherited categories become entangled and repurposed in new global logics of governance. They crystallize in new ways, becoming ambiguous, flexible icons that circulate through wider circuits of contemporary science power.

The pluripotency of the population is in that its singularity can be offered as a generality, and its fungibility is in how these categories can be made to travel over space, encompassing larger and larger swaths of a racialized humanity. For instance, A*STAR states that the major "strategic research thrust" at Biopolis is its development of "stratified medicine." The official claim is that "Singapore's multi-ethnic population of Chinese, Malays, and Indians is largely representative of the population in Asia, which makes up more than half of the world's population. Many pharmaceutical companies are increasingly viewing Asia as a major growth area, especially since there are a variety of diseases common in Asia . . . such as gastric and liver cancer, and various infectious diseases."[34] My research tracks the construction of the epistemological infrastructure and the reasoning and methods that underpin the creation of stratified medicine, which capitalizes on pluripotency by generating novel values out of linked data points on DNA and "Asian" elements that can be converted into patents as well as therapies. In the process, I illuminate that "races" are not immutable facts of the nature of the human species, but rather the ongoing achievement of complex biological, political, and epistemic processes. This book might be read as a way of telling a history of the present and the near future in which older forms of racialization interact and refunction within scientific endeavors, emerging markets, and the governing of security.

A Genomic Origami

In 2010, Liu summed up the goal of all the busy, mysterious work going on in the humming, dust-filtered, blindingly bright labs. "The Biopolis," he said to me, "is about making DNA fungible." He said that the information generated from DNA sequencing is fungible, providing an entry point to the bioeconomy because everything can be reduced to a sequence basis. The original usage of fungibility has both economic and legal components, in the linked notions of interchangeability and substitution, and of transferability as well. In economics, fungible assets would be commodities, options, and securities that are interchangeable and identical in value. By analogy, making DNA fungible suggests transposing (in data and abstract forms) qualities into equitable values. In other words, it is not the fixing of biodata, but the shifting around of data points that makes them innovative.

This way of using DNA data to generate transferable value is in contrast to "biovalue," which is an important concept that recognizes inclusion of biologically engineered vital qualities in the production of capital value.[35] More recently, Catherine Waldby and Robert Mitchell recognize that biovalue can be situated within a gift economy as well as a commercial one.[36] The notion of biovalue has been treated as a stable entity than can be activated in diverse domains (market, commons, and ethics). Instead of biovalue being a chameleon entity—now commodity, now commons, now affects—we might think of it as all these things in an interlocking information system that capitalizes on the aggregation of transferable science, cultural, and economic values. In novel research milieus like Biopolis, biovalue is a heterogeneous scientific object, its meaning always unstable and ambiguous when materialized as a signifying power that gathers up diverse local components bound up with health and wealth.

Perhaps "making DNA fungible" may be said to be inspired by a pluripotent reasoning that potentialities can come out of strategic recombination. Fungibility is engendered by using metrics of biological and social differentiation—DNA, mutations, biomarkers, ethnicities, and ethics—that render them equitable qualities in a single "Asian" system of biosocial values and valuation. In other words, here is a logico-semantic maneuver whereby all these disparate bits of information are aggregated in an ecology of information. The diverse data points must be transposed into a homogeneous language so that they can be reordered and translated from various models and algorithms and data sets to information libraries. Digitally interlinked, identifiable, equivalent assets are made to perform as both market and social/ethical values, and their transposable capacities enact a specific system of calculation and valuation that productively expands the present as a resource for socio-calculative action in the future.[37] With the digital means for capturing and relating materials, a new environment is created for making things fungible through their interconnections and fluidity across virtual and material worlds and virtual and material entities populating them. Singapore's research milieu is thereby branded as the site to shape and handle a new infrastructure of pharma diversity for bodies in Asia.

In the early twentieth century, the anthropologist Gregory Bateson proposed a new epistemology, which emerged from ecology and cybernetics theory. Information can be viewed as flexible ecology (homeostasis) or a system that regulates and corrects itself as it integrates information, basing their operational modus on distinctions/differences. Information, "a difference that makes a difference," comes "out of a context into a context," thereby generating a new difference or information. Recursively, any change within the ecology leads to further changes or reactions, thus generating a new system. Therefore, to think beyond the economic, the Batesonian approach would situate bioinformatics in a so-called informational ecology, or the ecology of an autopoetic or self-affecting/regulating system.[38] As it is made to absorb information (even techniques) not native to the system, bioinformatics have nonetheless brought the context of Asianness (bodies, health, and region) to the global knowledge economy. It should not be surprising therefore that in making a new system of biomedical information in the global context of Euro-American medical knowledge, researchers in Asia generate "Asian" distinctions that make a difference in the ecology of biosciences.

Drawing on the Deleuzian notion of "the fold,"[39] I invoke origami to refer to the folding of disparate systems of code—ethnicity, disease, geography, and market—into a fluid tissue of interconnected data points. For instance, there is the identification of "Asian" biomarkers for "Asian" types of cancers, co-related with spaces of infection, specialized cures, and intervention. Origami-like relations help to connect "fluid objects" (pathogens, animals, and people) to the "fluid spaces" of research, markets, contamination, and containment. As a novel intervention into the biopolitics of uncertainty, multiethnic medicine has sociopolitical implications for what it is to be biologically "Asian" and for what "Asia" is as a space of vulnerability and intervention in the world of life sciences.

"Datadiversity," Geoffrey Bowker notes, also layers in shared meanings of identity, body, and place, thus marking where affective resonance can be invoked.[40] The knitting together of disparate but identifiable assets in Asian genomics also animates productive affects of Asian identities. The invoking, provoking, and production of ethnically ordered science induce social conditions of being imperiled as ethnic and national collectivities, but are also being targeted for customized intervention. Diffuse affects of common endangerment and hope in turn support the building of this knowledge, stirring identification with a science brand that for the first time gives value to the specific afflictions associated with Asian peoples and their health needs as defined through this calculation and valuation of stratified medicine. In short, race/ethnicity is recast as both the mechanism of pluripotency and probability, the representational soul of the machine. As Michel Callon has argued, when the performativity of the market includes the overflow of affect and action, the simultaneous processes of commoditization and decommoditization are in play.[41] This genomic origami thus transmits various affects— genetic pride, public support—that Singaporean researchers hope to leverage for market competitiveness and science solidarity in the Asian region.

The Biopolis style of genomic science deploys race and ethnicity as *active,* not *reactive,* affects to induce potential values of solidarity and sociality. Such a scientific endeavor demands an attention to the generative, affective possibilities of scientific research, even when, and indeed *because,* it relies on long-standing categories of ethnicity. Because biomedical science in Asia—Singapore, Japan, South Korea, and China—is predominantly a state-funded and organized project, though with uneven degrees of regulation and marketization, racial and ethnic differences are absorbed as active variables and affective potentials: especially in the mapping and analysis of genetic variants, divergent molecular pathways, and expressions of diseases that vary across populations. Life-science practices thus draw upon and integrate ideas of racial and ethnic differences into their calculations, objects, and goals for populations already framed as a diversity of racial and ethnic groups. Furthermore, because scientific research is almost exclusively conducted in state-funded institutions, and often as a supplement to a deracinated (white) Euro-American human biology, scientists as public servants are socially bounded to serve "their" people and the use of ethnicity or nationality as active affects demands an alternate and even patriotic reproblematization. Communities of state-supported scientists are the instigators of social responsibility organized from the top down, fostering a contrastive model of biological citizenship than those that spring forth from grass-roots organizations.[42]

I am therefore not making a claim to an Asian science based in deep cultural features of the wide and diverse region known as Asia, but rather that new biosciences draw on and generate multiple ways of thinking and practicing Asia, which has always existed in relation with the world beyond. Two models of enigmatic DNA machines—one built by Biopolis in Singapore, the other by BGI Genomics in China—draw less from Asian medical traditions than from cosmopolitan science to realize variegated types of potentiality from a mixture of ethnic, economic, and scientific dreams. Nevertheless, as we shall see, the dazzling machinery of genomics and cutting-edge discourse is not devoid of the occasional glimmer of ancient beliefs about origins, bodies, and differences. My major focus is on Biopolis as a venture to extract pluripotent values by making genomics fungible and in the process calculating but also confronting a variety of risks that both enable and challenge the Biopolis initiative. A final chapter brings in BGI Genomics as an alternative use of ethnic data for genomic research.

In brief, the fungibility of DNA is created not by fixing biological coordinates but by aligning relations among bits of information in order to discover fungible aspects of variation. By holding their ethnic forms but shifting their relations, researchers at Biopolis hope to generate a spectrum of values that yield insights for customized medicine, to manipulate market risks, and to enhance affects of identity for scientific collaboration across the region. In its quest to map and enact a generalizable database for all of Asia, the Singaporean case of genomic origami illuminates how cosmopolitan bioscience is rife with topological possibilities.

Configuring an Ecosystem

Therefore, we need to situate the bioscience enterprise, and its operations and practices that enact and produce space-time configurations. John Law and Annemarie Mol argue that the double location of science and technology in labs and institutional networks implies that technoscience is "caught up in and enacts" the topological forms of "region" and "network." Science and technology also "exist in and help to enact" additional spatial forms that are fluid and constant objects.[43] Therefore, making DNA fungibility through strategic mapping, mobilization, and folding of diverse points and sites is productive of value-producing systems.

Throughout the book, I often refer to the "Biopolis ecosystem" or "Biopolis complex" in order to indicate that while Biopolis has its own campus one degree north of the equator, it is the center of a network of institutions, public and private, state and foreign, on a Singaporean island (see figure I.2). In other words, publicly funded projects at Biopolis are often connected in some way with these other institutions that may supply supplementary expertise, data, tissues, and critical funds. In addition, PIs from Biopolis may hold positions as professors in the national universities and hospitals, and corporations may become interested in their particular lines of investigation. Therefore, by the Biopolis ecosystem, I mean this network of collaborations and resource sharing that links four nodes of bioscience activities scattered across the island. The linked sites include Biopolis at One North; the main campus (major universities); "Hospital Hill," where public hospitals, clinics, and the Duke-NUS Graduate Medical School are located; and, to the west, a cluster of corporate manufacturing facilities near Changi International Airport.

Biopolis is a functioning knowledge ecosystem, with specialized niches and the circulation of actors, practices, and objects among them. Public research institutes, public universities, and hospitals are located near the downtown area, while global drug companies are at the periphery near the airport. Figure I.2 shows this interwoven public-corporate bioscience world, with dozens of corporate labs inhabiting Biopolis, American research programs embedded in national universities or hospitals, and the Duke-NUS Graduate Medical School on Hospital Hill. To put things too simply, state venture capital, public scientists, research material, and data are found in Biopolis, the

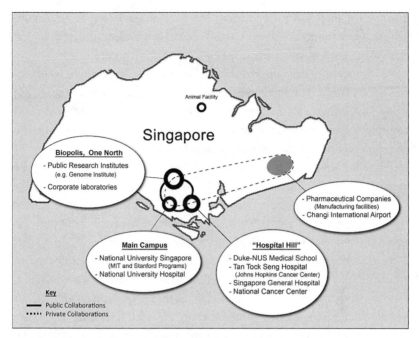

FIG. 1.2 The Biopolis ecosystem, Singapore. Diagram by Robert R. Ng.

main campus, and Hospital Hill. As a quite cohesive, state-dominated half of the ecosystem, they attract drug manufacturing companies looking for promising investment opportunities. In their quest to be "global," universities as well as corporations have been pulled to well-regulated Singapore in search of Asian bodies, diseases, and data for making novel medicine.

This book is interested in how situated cosmopolitan technoscience, by combining materialist and symbolic powers, exists in and enacts fungible spaces and objects that help shape a regional security. In taking this materialist-posthumanist stance, my goal is to illuminate as well how experiments are forged in conditions of uncertainty. This maelstrom of Asian bioscience is traced from the molecular to the cellular to the corporeal, and from the institutional to the communal, national, regional, and global scales.

Cascades of Uncertainty: An Outline

Contingency is modern society's defining attribute, and the very techniques to temper uncertainties engender more risks for which we are often unprepared. For Niklas Luhmann, experts increasingly operate in an "ecology of ignorance," a space of the unknown future where we deal with only probabilities or improbabilities.[44] Anthropological literature on governing future contingencies offers some insights on security threats as problems of technological governing. Andrew Lakoff and Stephen J. Collier have argued that twentieth-century modernity has required the development of "preparation technologies" that plan for potential "events whose probability is not calculable but whose consequences could be potentially catastrophic."[45] In this study, I consider a spectrum of modern techniques at multiple scales designed to shape a more secure future, but I concede that there are uncertainties—sprung upon us by the contingencies of politics and nature—that may be beyond technological control.

I recognize three figures of uncertainty that confront the Biopolis initiative, along a continuum of more or less calculability, amenability to preparation technologies, and anticipated but radical contingency.

Part I investigates how researchers in the Biopolis complex calculate an array of risks—genetic, disease, and ethnic—in building the knowledge infrastructure that underpins Biopolis's claim as the biomedical hub of Asia. We live in a "risk society," Ulrich Beck observes, in which scientists (and policy makers, hopefully) reflexively respond to uncertainty by devising risk calculations to counter a range of random threats.[46] Indeed, modern power has relied to a large extent on mathematical techniques as the basis of rational decisions for mastering uncertainty. For instance, the health sciences have moved from moralizing claims about good or bad, to mathematical calculations of normalities for managing life.[47] Some uncertainties are productive in that they are reducible to rational calculations and risk assessments. In *Security, Territory, Population*, Foucault traces the modern calculation of biological events—morbidity, mortality, and risk—as vital to governance, or the biopolitics of security.[48] Power over life is continually reorganized, and the interplay of diverse statistical normalities—the "law of large numbers" and the "stable" object to be measured[49]—permits the prediction and projection of collective risks.

Statistical devices have now become an important foundation of post-genomic science. Algorithmic formulations now extend the calculation of normalities and probabilities of risk to the molecular level. Risk genomics is in part driven by data diversity that is both foundational and performative of research, affective, and market values. For instance, by aligning variations in ethnic (Chinese, Indian, and Malay), genetic, and disease information, Biopolis hopes to make DNA fungible, thereby positioning itself as the site for research on majority populations in Asia. Of course, computations themselves tend to generate other kinds of risk, from the misfolding of data to perturbations that build into multiplier effects. The first three chapters explore how the ethnic heuristic is variously deployed: in biostatistical databases and in disease science, cancer research, and other programs that help the Biopolis ecosystem serve a broad "Asian" market in pharmacogenomics.

Chapter 1, "Where the Wild Genes Are," explores how the deployment of ethnic variables and heuristics operates, both as the artifact of a confluence of historical and epistemological conditions for cosmopolitan medical science and as a site through which new conceptions of Singaporean populations are articulated. I illuminate how the National Institutes of Health (NIH) policy of racialization-as-inclusion in research informs the building of Asian DNA databases at Biopolis. Singaporean biostatisticians maintain that genetic traits among populations in Asia that are relatively new to medical genomics gain value from being calculated and databased. Singapore's ethnic-specified DNA databases, scientists claim, are more "competitive" than those from Europe that lack such ethnic diacritics. I argue that ethnicity is rendered an immutable mobile that circulates databases beyond tiny Singapore, permitting tiny Singapore to represent an entire continent by shaping a topological space of biomedical "Asia."

Chapter 2, "An Atlas of Asian Diseases," gives an account of how Singapore drew on its biomedical resources in order to launch Biopolis as a site for clinical testing and medical tourism. Public-private joint ventures and the establishment of a bioethics committee quickly made Singapore a site for stem cell research and organ transplantation. The assembling of an atlas of Asian-type diseases became the foundation for biomedical research, as well as the mapping of "Chinese cancers." By thus differentiating from a "universal first world body," Biopolis assembles the data, information, experiments, and meanings of "Asian diseases," configuring a potential cancer research market as well as an economy of reciprocal research.

"Smoldering Fire," chapter 3, discusses the identification of genetic risks for forms of cancer prevalent among ethnic Chinese and other Asian groups. I explore in particular how cancer research at Biopolis, especially those projects that mine large amounts of racialized genomic information, creates the conditions for new ways of understanding and living in susceptible bodies. By identifying genetic and ethnic risks, cancer research engenders contrary affects of vulnerability and optimism. I discuss how researchers in Hong Kong and Singapore create novel objects such as the "Asian female nonsmoker" as biomarkers for certain cancers. The self-performance of the clinician-scientist illuminates how, as a boundary subject, he is poised between seemingly objective scientific work and the ethical promise of customized therapy. The search for cancer biomarkers engenders a state of attachment to a disease that comes to be imbued with both dread and hope.

Part II, "Uncertainties," covers those contingencies that cannot depend on quantitative risk calculations, but rather come to rely on preparation technologies and infrastructures that anticipate a range of economic and scientific challenges. Entrepreneurial uncertainties that Biopolis must contend with include shifting global conditions that impinge directly on the success or failure of the state-funded biomedical enterprise as well as the competitiveness of the infrastructural and experimental aspects of the project. International standards of bioethical experiments, including the establishment of internal review boards, are part of the arrangement necessary for the conduct of reputable science. Chapters in this section explore a variety of challenges confronting the Biopolis science initiative: the role of bioethics in the success of a biomedical enterprise; the uncertain meaning of scientific virtue in a milieu of expatriate scientists; uncertainty in funding levels; and promising outcomes for high-stakes experiments.

Capitalism deterritorializes all previous existing codes in order to become the universal coded form as capital.[50] In other words, volatility in market and knowledge flows can be mitigated by establishing codes that standardize biotechnological rules and practices that facilitate market flows. In the pharmaceutical industry, Andrew Lakoff observes, diagnostics technology must be in place to ensure "liquidity," or the capacity of information to acquire value through circulation.[51] In addition, the transition from medicine to biomedicine involves the recasting of medical architecture and infrastructure. Global competitiveness requires the building of the "biomedical platform," defined by Peter Keating and Albert Cambrosio as a specific configuration of instruments, individuals, and programs, an institutionalized space that generates routines, entities, and activities held together by standard reagents and protocols.[52]

In Singapore, the preexistence of legal and business regulations helps boost the capacity of its biomedical project to engage global business. The country's reputation—consistently ranked highly as one of the least corrupt places to do business—as a corporate and financial center helps to reduce some of the uncertainty of roiling global markets. The strategic mix of "best practices" in global business and cosmopolitan science has created as "risk-free" a zone for investing in science research as anywhere in Asia. Critically as well, widespread fluency in English and Mandarin Chinese, and multicultural experience in bridging Asia and the West, allows the city-state to be a matchmaker between scientists, data, samples, companies, and cultures from diverse sites. But such systems of value creation, we should not forget, are vulnerable to the vagaries of situated politics.

Will "small, smart, and nimble" Singapore indefinitely provide generous public support for the doing of cosmopolitan science? Singapore's engineers and economists control state funding, and they tend to be impatient, results-oriented leaders. How long will state managers waiting for findings "from bench-to-bedside" continue to support Biopolis as a biomedical hub in Asia? During his directorship of the Genomic Institute, Edison Liu's job was as a fervent rainmaker. In order to please his paymasters, he needed to justify spending on research by projecting long-term and short-term metrics of bio-science output. In his more informal moments, he envisioned at minimum four decades of state investment, amounting to some US$40 billion. After all, post-genomic science research needs a long period to make discoveries that can be proven valuable; state-supported science is especially critical when big pharma routinely avoids research that has no immediate market application. Uncertainty in funding is a constant for all scientific experiments.

Chapter 4, "The Productive Uncertainty of Bioethics," explores the theme of how bioethics and other regulatory regimes can reduce uncertainty surrounding the viability of a biomedical initiative. The chapter follows Asian researchers in their own working through of the limits and contradictions in a universalized ethical framework as it plays out in their various fields and sites. In contrast to a focus on bioethical violations in the emerging world, scientists in Southeast Asia view global bioethical regulations as inadequate in at least two ways. Bioethical guidelines such as "informed consent," they argue, are not able to address the substantive needs of indigenous donors. Second, the application of bioethics alone does not guarantee normative conditions that regulate any reputable biomedical science endeavor. Biopolis illuminates how bioethical procedures need to be embedded in a biomedical platform and facilitated by cross-cultural skills to deal effectively with international science actors and institutions.

Chapter 5, "Virtue and Expatriate Scientists," examines the unstable meaning of virtue in science as it goes global. It argues for a notion of "situated virtue" by exploring how a variety of researchers "collected" in Singapore negotiate the unstable meaning of virtue attached to the science enterprise. Superstar Western scientists are in Singapore to seek great working conditions, access novel data, and sometimes have a chance to do good in Asia. Foreign lab assistants, many from China and India, tend to view science as a lucrative job that gets them overseas. By contrast, for locally born scientists, scientific virtue and civic duty are entangled in the emergence of Singapore as a regional biomedical hub. The effect of *kiasu* is an additional pressure on native scientists to recruit, train, and inspire younger Asian scientists to eventually take over the enterprise and shoulder regional responsibilities.

Chapter 6, "Perturbing Life," explores the world of stem cell research, a high-stakes field of rapidly changing innovations that pose difficult technical and ethical challenges for developing immunology. The ethical debates over stem cell research in the United States, combined with historical strengths in livestock breeding in Asia, created an opportunity for the development of stem cell research as a distinctly Asian field. As researchers attempt to use stem cells for modeling diseases, they continue to be haunted by the question of whether and when iPS cells will ever be viable and useful for developing medications for autoimmune conditions. Experiments with iPS cell technologies also have larger implications for our changing notions of the cell, the Asian body, and the body politic. Finally, the prominence of researchers of Asian ancestries in cellular research worldwide has led to the view that it is an arena of "Asian" specialty and intra-"Asian" rivalry, thus adding yet another uncertain element to this highly competitive field.

Part III, "Known Unknowns," considers the challenges of meeting radical uncertainties that combine potentially disastrous events with a sheer variety of possible outcomes. In February 2002, Donald Rumsfeld, the former U.S. secretary of state, under questioning by the press, invoked "known unknowns": "that is to say we know there are some things we do not know."[53] I use "known unknowns" to consider how experts must be ready to take responsibility for any contingent outcomes. Because potentially disastrous events such as pandemics and climate change distort our temporal and physical coordinates, they are semilegible and defy conventional methods.[54] Therefore, security initiatives depend on "imaginative enactment" or scenario-building exercises, which are key ways in which possible future crises can be generated—and therefore prepared for—in the present.[55]

Cosmopolitan science confronts a dizzying array of interconnected possibilities engendered by shifting knowledge, contexts, and contingencies. Beyond the focus on "preparation" in the form of anticipatory enactment, the Biopolis and BGI Genomics cases include not only the management and mining of flows of populations, data, tissues, and other objects, but also the shaping of strategic international relationships and the reimagination of belonging and of Asia. Therefore, the scale of potential intervention is not always as clearly given as in a U.S.-focused understanding of biosecurity in a terrorism framework. Biothreats, or pandemics, in Southeast and East Asia are borderless and are perhaps more about an existential problem of living in a region with neighbors who may or may not cooperate. In post-SARS Asia, preparation technologies—public health interventions, genomic infrastructure, and disease surveillance systems—are being put into place, but political uncertainty remains as to whether different countries can come together as an epidemiological region in combating disease emergence. Uncertainty of cross-border coordination is ramified by the uncertainty of nature, in the form of newly emerging infectious diseases and potential disasters triggered by climate change. But as we shall see, Chinese scientists seem to view the future as a shifting mosaic of elevations and temperatures that will spatialize human habitation in ways that demand new arrangements of biogenetic capabilities. A known unknown is the kind of uncertainty surrounding the misalignment of the epidemiological and the political Asias. Related political and natural unknowns also haunt the future of cosmopolitan science itself. The final chapters consider the gap between the known and the unknown in anticipating high-stakes events. Policy makers and scientists are confounded by unknowns surrounding transborder science collaborations, capacities to deal with the next pandemic, and the health effects of climate change.

Chapter 7, "A Single Wave," discusses how Asian scientists interpret population genetic data in order to create a story about the conceptual unity of diverse peoples on the continent. Against the backdrop of historical and continuing political tensions, scientists at Biopolis have led the effort to form a first-ever trans-Asian genetic network. The assembled genetic data have permitted researchers to claim that a single human wave out of Africa populated the Asian continent, thus challenging an earlier anthropological model of a two-prong entry. By stirring affects of genetic pride, storytelling participates in a scientific renewal of "pan-Asianism" by getting disparate colleagues together in a single biomedical commons. Despite a new imaginary of a unified Asian past-present and a potentially collective present-future in science, it remains unpredictable whether deep trans-Asian factionalism can be overcome to confront future epidemiological threats in the region.

Chapter 8, "'Viruses Don't Carry Passports,'" discusses the rise of Singapore as a potential CDC-like center for a tropical region teeming with deadly viruses. In the aftermath of the SARS pandemic, the Duke-NUS Graduate

Medical School established a program to deal with epidemiological dangers that are still unknown, that is, the "newly emerging infectious diseases" that threaten the region and beyond. The chapter frames the battle against tropical diseases as an emerging biosecurity assemblage that shapes cascading scales of intervention. It identifies problems presented by the flows of "mutable mobiles"—deadly viruses, their animal and human carriers—as well as spatializing techniques from the molecular to the zoonotic to the national and global scales. In addition, international health, corporate, and U.S. military agencies are ready to be part of the assemblage in times of emergency.

The final chapter, "The 'Athlete Gene' in China's Future," shifts to South China, where BGI Genomics provides an important contrast to Biopolis in its mix of a commercial global thrust and the use of ethnicity in a national framing of genomic science. BGI has become "a global DNA assembly factory" for having sequenced most of the world's life forms. Domestically, BGI deploys official *minzu* categories that reinforce the national model of a Han majority versus non-Han minorities. A Tibetan-Han DNA study is focused on finding the "athlete gene" that may provide insights for developing therapies for Han people, who lack the physiological adaptation for living in oxygen-thin highlands. This preemptive focus on a biological capacitation of populations suggests that China's scheme of official ethnicities is conceptualized as a diversified pool of genetic resources for the fortification of China's genomes against the pressures of an environment to come. I illuminate how scientists at BGI are attuned to scenarios of catastrophic events associated with China's huge, aging population and the survival challenges of climate change.

The epilogue returns to the ethical quandaries of a technology that, by seeking a pluripotent fate, may indeed open us up to a multitude of "unknown unknowns." I compare Biopolis as a transborder biomedical zone that acts as a "DNA bridge" to American cosmopolitan science, to BGI Genomics as an octopus-like global biotech enterprise that also has a domestic agenda anticipating China's national health challenges. These contrasting modalities of Asian biomedical entrepreneurialism both particularize and universalize the life sciences as we know it. The pursuit of fortune, fungibility, and hope in bioscience, I conclude, must confront fear and the finitude of life itself.

PART I

RISKS

ONE

WHERE THE WILD GENES ARE

From 2004 to 2013, I explored DNA research in a biomedical frontier in Singapore called Biopolis. In 2010, I met Dr. Yang, a tall Singaporean whose vivacious personality belied the nerdy image of a biostatistician. He welcomed me in the manner of the bright young scientist on the cusp of something big. During his stint in the Singapore army, a requirement of all able-bodied male Singaporeans, with the unit of defense science, Yang became interested in "how genetics affect traits in Singapore." For instance, he said, "Ninety-nine percent of the Chinese here are myopic. Obesity [among ethnic Chinese and Malays] and diabetes [high rates among ethnic Indians] are the foci of defense science in the island-state." The security picture he referred to conjures the island as differentiated pools of genetic material and vulnerability. After his army service, Yang returned to Oxford University to work on a doctorate in biostatistics, when he was awarded a Wellcome Trust grant. Yang described this still-emerging field of biostatistics thusly: "There has been a logical progression in biology, physics, and chemistry from observation to math science. We deal with data quantitatively rather than deterministically; that is, the interaction of genes and environment. When risks are found, people will manage their health better."

Yang and his colleagues are incubating a new kind of biological science that originated in the West. They use genome-wide association studies (GWAS) to scan complete sets of the DNA of people in order to develop probability profiles of associations between ethnic groups and a disease under study. Studies also link data on genes and the lifestyles of large samples of ethnic groups in order to find statistical patterns of risk for certain diseases such as type 2 diabetes. As Yang's quote above discloses, objects in Singaporean genomics are not only identifiable health risks, but also ethnic categories central to administrative practices in many Asian countries. As a legacy of British colonial rule, statistics and racial and ethnic categories have long been involved in the biopolitical management of the health of the population, except that with the widespread deployment of biostatistics, larger geopolitical and epistemic ecologies are situated somewhat differently today. Because Anglophone ethnic categories find resonance in varied postcolonial administrations in Asia, they can escape their national contexts. Therefore, Asian racial categories have become heuristic devices of comparative value in the work of universalizing situated genomic science.

Toward an Inclusionary Genomics

The Human Genome Project, completed in 2003, drew on the genes of a few individuals to map a "universal" human genome, one that does not exist in the body of any specific person. U.S.-based scientists were interested in representing humanity in general, albeit dominated by genes from the maverick scientist-entrepreneur J. Craig Venter, whose company Celera Genomics (now Synthetic Genomics) had launched the sequencing of the human genome in 1998 but who decided to join forces with the National Institutes of Health (NIH).[1] The joint efforts of Celera and the NIH came together in the version of the Human Genome Project, predicated on a singular mapping of the human that was presented to the world. Paradoxically, this vision of generalized, deracinated genomic humanity seems to challenge an earlier NIH policy that promoted the racialization of human genomes.

In 1993, the NIH Revitalization Act established guidelines for "the inclusion of women and minorities as subjects in clinical research."[2] In contrast to the Human Genome Project, the NIH's earlier racialization-as-inclusion is part of some attempt at social justice through more representative data. In the United States, progressive scholars did not seem to have any issue with the universalized model of the human genome, but they did with the policy that sought to link racial and demographic information to human genomes. Social scientists have railed against the racial marking of genetic risks or biomarkers because of possible stigmatizing effects on minority subjects. However, the medical anthropologists Margaret Lock and Vinh-Kim Nguyen cautioned that the NIH used self-identifications of gender, race, ethnicity, or the preferred term "ancestry," not as discrete categories but "as heuristic devices for studying the frequency of specific genetic traits in 'at risk groups' that accompany the molecularization of race."[3]

From a materialist–posthumanist stance, I would note that while sociocultural anthropologists allergic to the biological continue to debate the relative effects of "nature versus nurture," biological scientists have gone ahead and culturally imbibed much of the critique while instrumentalizing it to differentiate the new racial medicine from older racial sciences. This quest to racialize human genetic fingerprints is the latest instance of how the biological sciences are increasingly engaged in the digital project of "making up people."[4] The life sciences are a cultural-scientific endeavor that increasingly relies on engineering and biostatistical logics, not old-fashioned racism, to frame questions of the biological.

Because the Human Genome Project was initially limited to sequencing DNA from only four regional groups, it set off a race to study human genetic variations across the world. Pharmacogenomics quickly became a global growth industry, creating huge demands for high volumes of data on genetic defects associated with deadly diseases for which drug companies sought to develop new therapies. Racialized genomes have become a technology of personalized medicine, providing clues for gauging genetic susceptibilities in certain populations. An American scientist working at Biopolis noted, "Most genomics research has been done on Caucasians based in Europe or the U.S., and we are only just starting to understand about how applicable these findings are to worldwide or Asian populations." From the Singaporean vantage point, therefore, hunting for DNA in the relatively unmapped human biological resources in Asia provides a major opportunity to be a player in pharmaceutical research. Thus, the NIH research practice of racialization-as-inclusion migrated to Singapore, but there the races/ethnicities included are not minorities but majorities in an emerging global region. Yang, our enthusiastic biostatistician, is a leader of the Singapore Genome Variation Project (SGVP), which Singaporean researchers readily compare with human genome diversity projects in the West.

This chapter argues that genetic traits among populations in Asia that were, up until recently, relatively new to medical genomics—and thus "in the wild"—gain value from being calculated and databased. First, I discuss how Singapore's technocratic rule and ambition to be a biomedical hub capitalize on its racially ordered medical data. Second, the state is the venture capitalist driving the hunt for "genes in the wild," so it adopts the American ethnic heuristics to make its database standardizable and mobile beyond the island through achieving a compatibility that makes Singaporean databases accrue value in a global environment. Third, in a bid to position an Asian database, researchers mark a general trend toward valuing genetic diversity rather than homogeneity as the principal feature of a good database. They claim that the Icelandic deCODE database promotes DNA homogeneity, while the Singaporean database celebrates Asia-wide DNA diversity. The next section elaborates the different mechanisms that deploy the ethnic heuristic, making it accumulate diverse peoples and places. By thus generating distinctiveness in Asian genetics, biostatisticians can respond to the following challenge: how can the mapping of DNA variants on an island gather up, as it were, an entire continent?

From Multiracial Order to Transethnic Genomics

Despite some skepticism, there is no seamless continuity between Singapore's racial politics and the development of Biopolis as a center of Asian genomics. The main contribution of official multiracialism is health data diversity that was computed in the run-up to dual state projects of promoting medical tourism (Singapore Medicine) and the Biopolis biomedical research initiative.

Singapore's multiracial model of government was inherited from British rule. As former joint British colonies, Singapore and Malaysia inherited but have transformed the British colonial model of the "plural economy" made up of peoples with diverse ancestries, cultures, and religions coexisting tensely in a nation.[5] After independence, the two former colonies were briefly linked as a single nation before splitting into separate countries in 1965. Each has an official multiracial order, with ethnic Malay-Muslims as officially first-class citizens lording over ethnic Chinese and Indians in Malaysia,[6] and ethnic Chinese enjoying political dominance over Indians and Malay-Muslims in Singapore.[7] In the island state, a system of "CIMO multiracialism" (Chinese, Indian, Malay-Muslim, Others) aims to balance the claims of different races in the nation while also codifying the national body as a composite of distinct ethnic populations. Since their split-up, both countries have been poised on a knife's edge of tense coexistence and competition, thus making collaborations, including in the science realm, very difficult.

In Singapore, the CIMO model orders each race according to its relative demographic size, maintained as the approximate proportions of Chinese (more than 76 percent), Malays (14 percent), Indians (8 percent), and Others (1–2 percent). This system of "ascribed races," Chua Beng Huat notes, is a "discursive and administrative simplification and homogenization of differences that inhere" within each racial silo.[8] Officially, citizen-subjects are expected to identify themselves within a racial category, as well as the state's construction of a race's particular cultural practices.

Another colonial legacy in Singapore is a "rule of experts,"[9] whereby a technocratic elite decides on policies for the common good, often described as a form of political authoritarianism or a paternalistic form of government. The recently deceased first prime minister of Singapore, Cambridge-educated Lee Kuan Yew, exercised a style of "state fatherhood" that occasionally invoked Confucianism to buttress social trust and loyalty to the intelligentsia. This authoritarian multiracial order had in the past compared the "robust migrant stock" and "Confucian" values of ethnic Chinese to the "soft society" of native Muslim Malays.[10] The differentiated ranking between ethnic Chinese and ethnic Malays (with ethnic Indians somewhere in the middle) is evident in many continuing institutional and normative forms. The educational system is stacked in favor of the children of professional, Mandarin-tested, English-speaking ethnic Chinese. The meritocracy tends to promote ethnic Chinese as the brains shaping the nation and its future.

Perhaps not unexpectedly, the cultural superiority of being ethnic Chinese is haunted by a Singaporean syndrome "fear of losing out" (or *kiasu*, in Hokkien dialect) for many of the island's Chinese. This fear of failure is affectively connected to the risks of being a tiny Chinese-dominated island surrounded by a vast Islam-dominated Malay archipelago. *Kiasu* is manifested in everyday anxieties associated with school exams and sports, shopping and lifestyle practices, as well as in official warnings about potential racial strife and even regional conflict. Administrative multiracialism predicated on risks of losing out recursively legitimizes authoritarian rule for solving problems of national well-being and security by lifting the state fortune beyond its geographic isolation. The state carrier Singapore Airlines was a first postindustrial flight beyond the island's location in Southeast Asia, as were bids to transform Singapore into a corporate and financial hub of Asia. Making Singapore a knowledge economy is the next leap onto the global stage.

By the turn of the century, the technocrats sensed that local politics of official multiracialism under the umbrella of a single nationhood was too constraining for a city that wants to be truly cosmopolitan and globally significant. Loosening the grip of multiracialism, Singaporeans began to enjoy greater social latitude and recognition of cultural identities and expressions linked to ancestry, place-origin, language, and religion, as well as sexual orientation.[11] Singapore soon became an upscale city for the global rich, and its shops, restaurants, nightclubs, residences, museums, and lifestyle scandals were featured in international media. Currently, one quarter of the more than 5.5 million residents is foreign. Against the state's insistence on "racializing" everyone, academics and ordinary Singaporeans often use "race" and "ethnicity" interchangeably, to emphasize the voluntary, fuzzy, and performative aspects of social identity. The question is, will the use of the ethnic heuristic in biomedicine reinforce a colonial sense of racial biology?

This burst of globalized cosmopolitanism, however, is not the main reason why ethnicity, not race, is used as a marker in Singapore's genomic science. Rather, researchers in Singapore adopt the American model of the ethnoracial ("ancestry") genomic project, deploying the ethnic heuristic in the interests of an Asia-oriented inclusionary genomics. The state's medical establishment routinely collects anonymized and racially coded medical data from patients, which transition easily into data in research databases where demographics are already marked. Official political racial designations become a basis for classification of data by ethnicities, which make multiracial populations in Singapore stand as a microcosm of Asian ethnicities far beyond the island. By shifting from Singapore "races" to trans-Asia "ethnicities," but basically preserving the continuity of racialized medical data, the ethnic heuristic is widely used to track genetic variants and find mutations and disease risks among majority populations in Asia.

Two observations follow: the instrumentalization of biostatistical logic, and the need to subvert the purity of the truth metric typically associated with scientific points of view. First, biostatistics is now more intertwined with computer engineering logics, thus enhancing its entry into the global fray. By accommodating the Asianness of the ethnically heuristic data to a global bio-ecosystem of knowledges, biostatistics makes what it is doing intelligible to metropolitan sciences and pharmaceutical interests in multiple registers. Despite state-based funding for bioscience, the Biopolis modeling of multi-ethnic DNA is, first, a pragmatic strategy to produce a transnational statistical infrastructure that can be mined for its demographic and geographical traits by researchers and drug companies, and, second, a biopolitical investment in the health needs of an aging citizenry that can be managed through this circuitous route.

Second, the ethnic heuristic though, in the end, is itself a transposition or a sorting mechanism, with a normative rationality of its own, from the pre– and post–World War II European and American scientific contexts. So it is the inclusive logic of this ethnic heuristic that governs its possible lines of flight, taking shape in its modification and usage in the Singapore system, which in turn is more amenable and intelligible to the gatekeepers of the market they wish to co-opt. The commercial transnational thrust is greatly facilitated by English as the language of cosmopolitan science and its normalization of "ethnicity" as a politically correct term for correlations with genetic differences. By deploying the biomedical use of ethnoracial or ethnic difference as a heuristic in DNA research, Singaporean scientists can circulate their data and applications to sites where English-identified like-ethnicities are found. For instance, the English term "ethnic Chinese" in Singapore can be used without scientific awkwardness to refer equally to Chinese populations in Singapore and to certain groups in mainland China, across Southeast Asia, or in the United States, for that matter. By embedding "Asian" ethnicities in biostatistics, Singaporean researchers make genomics fungible by enhancing the capabilities of the data to be mobile, standardized, and marketable.

Therefore, the utility of the English term "ethnic Chinese" and its transversal heuristic applicability to a different class of objects uncomfortably goes back to the anthropologist regarding our own routine use of such Anglophone categories. Despite our protestations about differences and situatedness, do not anthropologists on occasion stretch "ethnic Chinese" labels to cover groups in different places? Have the scientists merely copied from our playbook? What is at stake when scientists use "our category" but apply it to other domains and situations?

"Gather Up as Much Information as You Can"

Modern knowledge captures, calculates, and invests in the multiple, variable, and unpredictable flows of things in the world. Thus, to discover unknown DNA variants, scientists need to look at genes "in nature," or "in the wild," or in places still outside the realm of calculations.[12] In biostatistical research, the axis of comparison is between the normal and the wild. For database builders, this "wildness" is less a place than a condition of the genomes of populations that is nonstandard for being outside normative databases.[13] Because databases are by definition comparative, the uncataloged genomic diversity of populations in Asia can be made to yield new information that challenges the normativity of existing databases elsewhere. It is in this case that both Asia, relatively unrepresented in existing databases, and diversity, in a biostatistical environment that has heretofore privileged homogeneity, fall into this wild.

By databasing DNA in a scientific frontier, scientists in Singapore hope to achieve a comparative advantage over already existing Euro-American genetic databases that have been dominant in the worlds of science and pharmaceutical research and development. Thus, "in the wild" does not only mean outside the lab, or not databased, but in general, it also means outside Western databases, even if the genes from Asian peoples are also increasingly databased. By finding genes in the wild, scientists map into existence a new biomedical resource with its own genetic databases, probability measures, and market potentials. Researchers building a new DNA database capitalize on potentially productive correlations of genetic and social variables.

A novel DNA database is thus a technology of potentiality, one capable not only of producing novel research values for drug discovery, but also of absorbing affective values surrounding "our bodies, ourselves" in racial, ethnic, geographic, or disease terms. By choosing to correlate "ethnicity" with genetic variation, researchers unleash the productive potential of an "Asian" ethnic-diversified database that, through the use of flexible scales, can represent various groups across a vast terrain. Here is a design platform for "genome geography"[14] that deploys elastic notions of ethnicity and scale.

The anthropologist Christopher Kelty observes that information technologies do not merely connect existing groups; "they generate the conditions of possibility for new collectivities—maybe even new kinds of collectivity."[15] Similarly, I maintain that Singapore's ethnic-DNA aggregation produces novel biomedical collectivities that are ethnic-associated, such as "Malay" or "Indian" or "Chinese," pointing to transboundary ethnic collectivities that may come to social salience as the demographic loci of new affects that can be formative of what Paul Rabinow calls "biosociality."[16] In addition, the ethnic, variable DNA data generate new forms of difference within and between biomedical collectivities that are defined, sampled, and analyzed. In turn, the designation of ethnic-correlated biomedical collectivities also engenders novel notions of ethnicity, linking, for instance, "Indians" in Singapore with "Indians" elsewhere. Such ethnic, variable DNA objects, have, in the words of the sociologist Bruno Latour, "the properties of being mobile but also immutable, presentable, readable and combinable with one another."[17] An ethnic-DNA correlated database is coded to and implicitly indexical of broad racial-national categories that stretch over a broad and dispersed swath of "Asia."

The accumulative repetitions of the ethnic categories in medical records and databases make them both immutable and mobile. Through reference to shared co-ethnicity that stretches across merely political borders, an ethnic-specified DNA database can bring dispersed populations together. Such a biostatistical model is offered as a concentration and condensation of populations that make up a wide swath of "Asian" genomes, thus providing a new biomedical resource with regional reach. By thus correlating scientific and social variables and deploying their numbers, figures, and scales, Singaporean scientists have designed a DNA matrix that gathers up a huge and heterogeneous continent within this tiny island. Let us see how it happens.

Dr. Williams, a scientist who grew up in the New York metropolitan area, still seemed slightly displaced in tropical Singapore. In April 2010, Williams talked excitedly about Singapore's "electronic research habitat." Invoking Venter, Dr. Williams remarked that the new method in biology is to "gather up as much information as you can; there are no a priori right and wrong answers." Computational technology is producing a new way of seeing more and differently; the significant principle is between what is already known and what is still unknown. The goal of genomic sequencing is to unravel such information so that "we can come up with better interventions to sustain life."

Modern biology, Williams explained, "is all about automated machines churning out huge amounts of data, which then have to be stored, analyzed, and visualized. . . . Digital computing is the servant of nondigital, brain-based computing." In other words, genomic research seeks to bring order to huge amounts of informaticized DNA, establishing ethnic-risk-disease associations as a predictive tool or diagnostic screen that will help researchers to study cellular processes like gene function and metabolism. The goal is to find predictive "biomarkers" that link genetic defects to ethnic differences, disease susceptibilities, and prognoses. Computer readouts of genetic variants flag genetic susceptibility and ethnic association with a specific disease, helping clinicians decide on a potentially effective match of a patient with a particular drug. Genetic data do not provide a cure but offer a strategy of disease diagnostics that works closely with molecular research. Big genomic data are foundational to experiments that put into play a synergy between DNA defects and disease pathways, computer labs and wet labs, and biostatisticians and biologists. It is an integrative science that is under way in Singapore's modeling of a genomic "paradigm shift." The hunt for novel ethnic genomics in Asia, he remarked, was part of a global biomedical shift comparable to a model of President Nixon's war on cancer.[18]

How does Singapore position itself as an ideal site for waging the battle to accumulate health data and cure health threats in Asia? Singapore has an evolving public-private partnership in health care financing and provision, for the total population. The government subsidizes health care (along with housing and education) to make things affordable for citizens, but citizens share in the costs of the services they consume. Under the philosophy of shared responsibility (called the 3M system), Medisave is a compulsory state-run medical savings program for the working population. In 2005, it had accumulated an impressive S$30 billion (approximately US$1.8 billion) for future health care expenditure.[19] Public health care is provided by eight hospitals and five specialty centers that together account for 80 percent of inpatient beds. The state spends 3–4 percent of its GDP on health costs, achieving first world–class standards of health attainment. The government remains the biopolitical guarantor of medical service, despite a growing private sector.[20]

The Ministry of Health (MOH) centrally manages health care institutions to ensure that they provide efficient delivery and improved service levels. The MOH oversees public hospitals and controls health information at the national level. This produces a comprehensive collection of medical data gathered from patient records from birth and categorized according to four races. Because all Singaporeans, as users of health services, are automatically enrolled as genomic data points, the supply of genomic information corresponds *exactly* to the country's multiethnic citizenry. There is thus an orchestrated correspondence in different state agencies and on multiple registers between medical information, ethnic identity, and citizenship: something rather easier to achieve in an island city-state the size of a large American city run like a corporation than in the continental United States where the Affordable Care Act ("Obamacare") has been roiled by partisan politics.

Besides the centralized collection and computation of medical data, there is a coordinated network of institutions that controls and accesses public health information. The Biomedical Research Council oversees all public-sector research. Public hospitals, the National Environmental Agency, the Defense Science Organization, and the Genome Institute of Singapore come together to share expertise, resources, and strengthen research in the Biopolis ecosystem.

At Biopolis, as the leader in bioinformatics, Williams's role was to set up a computational grid for comparative genomics that draws on the DNA diversity of samples in and through Singapore. This is done by mobilizing and combining data on population genomics and medical genomics, all organized along ethnic lines. Within the next decade, electronic medical records in the island's public hospitals will be made available for data mining to foster medical research. This integration of hardware and software infrastructures aims to manage digital storage and flows between the island and other places in

Asia. In authoritarian Singapore, this hybrid computational architecture enhances conditions for work experience while also satisfying regulatory and legal requirements. Williams predicts that this "secure, scalable, and robust" genomics enterprise is part of the strategic building of Singapore's knowledge economy.[21]

This infrastructure in the Biopolis ecosystem, a public core with private elements growing around it aimed at creating an incubator for biomedicine, is very different from the construction of DNA databases elsewhere. The nature of Singapore's data, which is to say the nature of its multiethnic population and the cataloging of its alluringly scalable genome, is thus central in elaborating the comparative advantage of Singaporean science. A strategic mix of DNA heterogeneity and authoritarian politics gives Singapore leverage as a potential biobank for much of Asia. From their location in Asia, home to a diversity of peoples and nations, researchers at Biopolis think their genomic enterprise has more value than those in Europe.

Why Singapore Trumps Iceland

At the turn of the century, there were a few genomic institutions or companies, and they represented only a small range of human genetic differences in the world. The Hapmap is a haplotype catalog of variant genes that provides a shortcut to the inheritance patterns of DNA mutations. In a follow-up to the Human Genome Project, the Hapmap has expanded its analysis of the genomes of people from four to eleven groups,[22] but, as Singaporean researchers have noted, it covers only 5 percent of the world's population. They also pointed out that Iceland's deCODE Genetics is a company focused on disease gene mapping for only Caucasian populations. By contrast, in 2009, Biopolis led a fourteen-country initiative to collect genetic variants, SNPs (single nucleotide polymorphisms) under the umbrella of the Pan-Asian SNP Consortium. This Asia-wide database, Dr. Yang remarked, "is a great improvement" over the Hapmap and deCODE, because Asia (*and* Africa) are "where the rare variants are." As the development of pharmaceuticals shifts beyond the North Atlantic, diversity in human DNA data is a crucial resource. Scientists at Biopolis think of themselves as having a comparative research advantage, because Asian genetic variants are more valuable both to Asian scientists creating a new frontier in the new genomics and to drug companies developing new drugs and new markets in this populous region.

In the West, Iceland and deCODE Genetics, the company with exclusive rights to develop a comprehensive database on Icelanders, were hailed for the quality of its medical records and the sense of civic virtue that informed collaborations among the state and the academy. Gisli Palsson and Paul Rabinow suggest that "just as India is the official site for caste, Iceland is emerging as *the* site of biotech and bioethics."[23] But as genomic science spreads unevenly across the world, triumphant European models of genomics are less likely than once thought. The assemblage of genomic science and Singapore's differentiated medical data, as well as its location in Asia, make for a different design of genomic research and information.

Because deCODE Genetics focused on disease gene mapping for only Caucasian populations, Singaporean scientists therefore found it ironic that when deCODE Genetics made its initial public offering (IPO) on NASDAQ, it sold itself on the genetic homogeneity of its database. The anthropologist Michael Fortun reports that to set deCODE apart from the rest of the genomic companies, the company had to convince American investors that "there's money to be made from Iceland's genetic purity." But a commentator on SmartMoney .com raised objections, in an inimitable New Yorker lingo: "OK, it may be true that Icelanders don't all look alike. But that doesn't mean you'd pick Reykjavik as the setting for a documentary called 'People of Color,' either."[24] In deCODE's IPO, American investors were already apprised of the need to have genetic variability in databases. In the brutal pharma markets, what were formerly selling points for deCODE's scientific acumen and justificatory appeals to its ethicality were now increasingly problems for business modeling.

Besides its overly homogeneous data, Yang observed, deCODE is "too upstream" in its data formation to be competitive for biotech investors. Another limitation was the propriety controls that Icelandic citizens retain over their records, which limit diverse uses of medical data. In sum, where the homogeneity of Iceland's population's value depended on the creation of a data set defined by its internal consistency, Singaporean and Asian scientists argue precisely the opposite: homogeneity is not an asset but an impediment to robust data, and the heterogeneous composition of Singapore's multiple ethnic genetic pools is generative of its values, present and future.

In that global race, the richness of the data, the ethics of its management, and global marketability are all in play. Singaporean researchers note that the Biopolis databases, which integrate data from across Asia, match or surpass deCODE in each of these arenas. The Singapore database boasts representation of the three major Asian races, each of which exhibits tremendous internal diversity. This diversity, already categorized by racial cum ethnic data markers, makes it more robust in their terms. Beside this, the Singaporean assertion would be that not only do they have a more variable and therefore marketable ethnogenomic research science, they also have their own version of genomic ethics conditioned by concerns for collective rather than individual proprietary interests. Additionally, the official view is that Singapore's genetic diversity has appeal for global firms investigating new therapeutic possibilities in Asia.

But the Singapore-Iceland comparison also reveals the volatility of "value" in genomic business and how ethical, business, and promissory health values can converge and diverge in a fluid biomedical world. Despite its emphasis on ethical standards reflecting European values of freedom and privacy, deCODE floundered in the unruly markets for drug development. Meanwhile the kind of triumph of "value" in diversity that Singaporean scientists extol also points to the elusiveness of the anticipated creation of conjoined health: pharmaceutical, capitalist, and social values that promise to manage uncertainties in capitalist, health, and political futures. But Singaporean scientists like to think that they are thinking ahead by designing technologies that transform the promissory quality of genomic research and the market that sustains it into shared interests and fate for a region.

"Region," as a transethnic continuum, is emergent in these research programs. Clearly, as cells derived from different ethnic groups are brought under the microscope, the term "Asian" itself is becoming very elastic, referring, depending on context, to the genetic heterogeneity of the three major ethnicities in the region. The convergence of codes for DNA, ethnicity, and ancestral environments produces a mobile set of connections of scientific signification.

As Yang noted, "Our leverage is the multi-ethnic demography and the way we combine genetics sciences and traditional epidemiology." His genome variation project is based on assays of two million genomes in three Asian ethnic groups. Because of aging populations, aging drugs, and rising costs of drug development in the West, the moment is ripe for the growth of health markets in Asia.[25] Even though new drugs for diseases prevalent in the region are still many years away, I was told Singapore is making a head start by assembling DNA information that creates potential values beyond the island. Singapore's demographic diversity is thus offered as a pool of genetic assets in an experimental infrastructure concerned with variation over homogeneity, and the recodification of biological variability and mutability. The ethnic heuristic underpins Singapore as a research platform for all of Asia, both as a hub of biomedical expertise and data and as a scalable genetic microcosm of the vast continent's populations.

Ethnic but Anonymized Data

After all, the state plays a major role in financing and organizing health care, thus shaping an implicit social contract with the public to be supportive of the biomedical research enterprise. Perhaps more critically, the severe acute respiratory syndrome (SARS) epidemic reinforced a public consciousness of communal medicine, and it dispersed lingering doubts about the biomedical initiative that focuses on the category of "Asian" diseases.[26] After the SARS experience, journalists with the *Straits Times* explained to me why there is little resistance to the mining of medical data. "The public perception is that donations are to the public sector, that is, government hospitals and clinics which have 'respectability,' and citizens are absolutely willing to give samples for a public group, especially when the request is framed in terms of benefits to ethnic groups. A friend's parents, middle-class individuals, voluntarily donated their blood in the interest of medical science that may benefit Indians." The public blood bank is presented as a "national life resource," and parents to be are urged to donate their newborn's cord blood in order to ensure that leukemia patients in Asia will have access to "life-saving" stem cell matching.[27] At the same time, a growing private sector for cord blood banking is supplementary to the public blood bank, in part because of the understanding that leukemia patients are best matched with donors of the same ethnicity.[28]

By comparison, say, to Iceland, Singapore has a public system for aggregating medical data and tissues that is not hampered by proprietary challenges from patients and donors because ownership of data and tissues has been transferred to public institutions. Medical review boards in hospitals and research institutions have the freedom to determine the research uses of DNA data and tissues. In other words, the ethnic accrual of DNA variability is not merely economic, it is also productive of collective legitimacy. After all, producing DNA connections along lines of identity marked a form of identifiability for scientists. In addition, for donors and patients (potential and current), ethnic-identified DNA produces consent. For instance, patients in public hospitals routinely sign off permission for the anonymized use of their collected or discarded tissues for "scientific research," proliferating a distinctive form of life value.

Under the MOH, the electronic infrastructure gathers up all patients and makes them participate in a vast ongoing clinical trial of potential health problems. That is, public health care has as its condition the use of patients' data, which are "owned" by public health institutions. "Best practices" govern the gathering of new samples; patients sign consent forms, and patient information is converted into data points. In other words, the ethnic accrual of DNA variability is not merely economic, but also productive of collective legitimacy. Indeed, the researchers consider themselves to be engaged in a form of civic virtue by designing ethnic-differentiated DNA databases that are culturally identified with their "own" communities. However, some scholars worry that the use of actuarial risks linked to medical data may lead to racial stigmatization and other kinds of misuse of genetic information.[29]

A fundamental difference from the race and medicine debate in the United States therefore is that, in Singapore, people do not fear stigmatization by ethnic profiling but actually welcome medical research that targets their ethnoracial group for its promise of developing novel therapies. After all, people in Asia have not been historically included in biomedical research, and they welcome attention attuned to their genetic differences. There is also the concern that such predictive data on risk trajectory may influence insurance risk decisions that discriminate against people with preexisting health conditions. However, while predictive disease risks may become aligned with the life insurance market, the politics of health care in Singapore help to protect this kind of massive access to private patient information in its own way. Electronic health records are "de-identified" except for self-identified ethnicity. In order to further defend patients against the misuse of such records, the Singapore government plans to pass a law that requires medical insurance (Medishield) for all, regardless of any medical condition. By building on this robust infrastructure of health governance, the government makes it easy to access anonymized medical records, family histories, and disease risks aligned with ethnicity as the foundation for building "a valuable Asian DNA biobank in Singapore."

Ethnicity as Immutable Mobile

As an anthropologist, I wondered whether, in the attempt to come up with a general ethnic profile of risks, individual racial differences would be washed over. I was therefore disconcerted when Dr. Wu, a bespectacled, gray-haired, but youthful-looking geneticist, argued for a "bar-code" vision of ethnicity. He brushed aside my worries about the mapping of cultural and social categories onto cellular material as irrational and obstructionist in the urgent task of pursuing cures for Asians. (Scientists in Asia and the West alike deploy ethnic terms with the same aplomb that never fails to amaze anthropologists.)

It was routine, he said, for donors and patients to self-identify their ethnicity. Given the well-documented lives of Singaporeans, I suspected that researchers often used a mix of personal identity (ID) card information and medical records to construct the ethnic profile. "But what about persons of mixed parentage?" I asked. Wu impatiently noted that in such cases, as a matter of "practicality or convenience," as well as patrilineal bias, I may add, they used the father's self-identified ethnicity. Their aim was to obtain the general ethnic profile, not to be distracted by specific individual differences. He stated, "It is a matter of what resolution you want, or what scale in your sample to produce a reference database that can be used by researchers to trace disease prevalence." He continued earnestly: "The point is to develop a bar code that defines your ethnicity. Our final goal is to arrive at a gene that causes disease susceptibility, to finger that gene and pinpoint it." Thus the bar code reconfigures ethnicity in a set of statistically determined vulnerabilities that are linked to ethnic data populations. Wu also explained that variable DNA profiles exist in different geographical areas.

An epigenetic rule of gene–culture interaction correlates groups evolving in relative isolation with different kinds of susceptibility genes for certain diseases. He gave the example of malaria tolerance in some African groups, a microevolutionary outcome of what he called "in situ adaptation" that is associated with one or two characteristic genes. He went on: "Genetic pools vary in different places because they become molded by diseases prevalent there. Genetic features may account for resistance, so we are interested in finding that gene to develop a cure. Sometimes the [epigenetic] conditions that affect the vulnerable group are also taken into account."

In a post–Human Genome Project world, Wu emphasized, there was potential value in using the ethnic heuristic and "Asian" angle. He had been trained in Europe and been a visiting scientist in the United States and Japan. In the United States, he said, "They classified all Orientals together."[30] His point was that in Singapore and Asia, where larger-scale samples were more easily available, scientists could statistically stabilize the population samples to show "dramatic differences" among Asian races. These were categories with serious statistical amplitude. "There are huge numbers involved in our three representative populations: 1.2 billion Chinese (mixture of South and North Chinese); 1 billion Indians; three-quarters of a billion Indo-Malays, that is, almost half the world! These are considered distinct genetic pools." Through the use of ethnic-differentiated data, he seemed to suggest, a geometrical dynamic could be unleashed that expands the value of the data to staggering dimensions.

Yang told me rather boldly that the Singapore genomic data "trace differences and similarities among Malays, Chinese, and Indians, that is, races that represent one third of the world's population." Recently, leading hospitals, clinics, and labs on the island came together for a cohort study of gene–environment interactions in disease development among the three ethnic groups. The authors predict that "information obtained from the study could be applicable to India, China, and much of South East Asia."[31] The slippage from ethnogenome identity to ethnonations is very telling, for suddenly genomic and disease information assembled in Singapore has the potential to be biomedically relevant to populations in big Asian countries. How is that scientifically feasible?

The ethnography of scientific practices, Latour argues, reveals the transformation of lab findings into inscriptions: cascades of columns, diagrams, drawings, formulae, maps, and digital images that are combinational and mobile while remaining consistent as an optical power.[32] A useful analog is money, which circulates yet remains calculable and combinational. Ethnic bar coding of DNA develops elastic properties of the ethnic figure, to condense or stretch across sites, or to move without distortion (i.e., an immutable mobile). The repetition and displacement of the ethnic figures—Chinese, Indian, Malay—flatten their differences and permit the domination of the scientific diagram to do its work, at different scales.

The Ethnic Chinese Heuristic

How is this zooming in and zooming out of DNA data enabled by the use of ethnic heuristics? Ethnicity not only becomes a marker of genetic difference, but also functions as a biomedical category that can be flexibly applied to disparate groups in transnational space. But not all ethnic markers are of equal value at this moment in the making of a multiethnic DNA database intended to function in the region. The priority given to ethnic Chinese-marked genetic variants and disease susceptibility perhaps reflects the political bias of mainly ethnic Chinese scientists, as well as the desire to earn the support of the ethnic Chinese majority for the Biopolis enterprise. While Malay and Indian traits are part of Singapore's stratified medicine, there was the question of ease of gathering larger bodies of data. A scientist complained somewhat bitterly about the mistrust between Malaysian and Singaporean medical establishments that prevented a fruitful alliance around building up Malay medical infrastructure at Biopolis. Understandably, Malaysian academics and researchers wanted to control the sequencing of DNA obtained from their own populations, and they would not consider letting Singaporeans manage such precious racial patrimony. Deeper collaborations with colleagues in India were uncertain and the assumption was that Indian scientists would want to control the DNA information for their gigantic population.

The focus on ethnic Chinese genetic and disease traits is mainly strategic, driven by market and collaborative reasons. Biopolis aims to design a biostatistical infrastructure that can be directly applicable to the huge ethnic Chinese market in the region. Researchers point to the deadly diseases that threaten millions of Chinese each year, especially those with a genetic or metabolic basis. There is also the possibility for Mandarin-speaking Singaporean scientists to collaborate with colleagues in the People's Republic of China (PRC). The leveraging heft of the "Chinese" figure can code for DNA variants and other findings obtained from far-flung sites.

Ethnic Chinese DNA and disease risks are especially critical in building Biopolis up as a center for "Asian" cancers. The head of a cancer center in Singapore expressed a qualified statement about the ethnic heuristic in cancer research. "Phenotype can be described, but it is not clear that genotype is ethnically marked. Our goal is to refine existing treatment, to glean information for refined or novel treatment. This is an expectation; it is not clear whether it can be realized. But individuals do respond to different treatment." Nevertheless, the place to start looking for links between genetic variants and the incidence of cancers in Asian populations seems to be Chinese populations, preferably in China.

So I chased down the Chinese trail to the Genomic Institute of Singapore. There, I met Dr. Lin, a PRC-born oncologist who told me his project was on "how genetic variation is distributed in Chinese populations." To my query about the focus on Chinese groups, Lin replied, "I have very practical reasons for having an interest in China. The population is there. We need a lot of patients, that is, thousands of disease phenotypes. China is a major source of biomedical data [for our research here in Singapore]. We combine local and Chinese [PRC] samples." Expressing an elastic sense of scale afforded by computational biology, he noted, "They are all Chinese in a sense." This "in a sense" speaks to a strategic accumulation of individual phenotypes into variations of an overarching ethnic category distributed across space and national lines.

Using the Chinese bar code, Lin was able to accumulate far-flung allies and resources in one place at Biopolis. Being China-born, he easily forged links with many clinics and hospitals in China that supply him with the germ-line cells for different cancers. By integrating data and samples from China and overseas sites, he has built a huge ethnic-correlated database that transcends borders. Furthermore, the Chinese-labeled data can jump scales by becoming a paradigmatic form for "Asian" types of cancer. This projection and prognosis have to do with the frequency of cancers among Chinese populations, the accessibility and scale of DNA samples, and the multiplicity of environments in which Chinese peoples are distributed. The pragmatics of scientific research and design thus display how ethnic categories can slip, expand, and contract; and in this case, the relative nondetermination of "Chinese" can be wrought into a proxy for "Chinese" everywhere and, at times, even for "Asia."

The use of Anglophone ethnic names enhances the geometric power of DNA databases. English terms for different Asian ethnicities in Singapore (a legacy of colonial times) became the starting categories for indicators of DNA differences, so that particular ethnicities designated by the English language become aligned with differences in DNA, mutations, and disease expressions. Asian countries to which Anglophone terms refer are not only multiracial but have subethnicities, which become relevant or not depending on the way the data are divided—north and south Chinese, Sikh and Tamil, as different "Indian" groups, and so on, not to mention entrenched forms of social stratification and differentiation. Thus the mix of scientific artifacts and English terms engenders a series of cascading data that can transport and transfer the implications of DNA knowledge. By thus accumulating scales and flattening diverse populations across Asia, Singaporean genomics demonstrate that, unlike Iceland, no island is an island.

An Asian Genetic Architecture

The island has become a center of prognosis when it comes to ethnogenomic identities in the region. Dr. Tai spells out the implications within the context of the Biopolis hub. "Our databank represents a much more diverse population that is reflective of what will happen in much of Asia. . . . There are few places in the world where you can look at the effects of rapid socioeconomic development on three different ethnic groups that seem to respond somewhat differently to the environment. In addition, the rather good infrastructure and communications will give us advantages over other biobanks." Here Tai is interpreting different genetic responses to changing environmental conditions, which presumably include the social effects of structural inequalities that are ideologically washed over.

By mobilizing many resources—Asian genetic diversity, global expertise, and regulatory governance—Singapore projects itself as a prime center that links major ethnic collectivities to risk diagnosis, prognosis, and drug discovery. Williams predicted that "authoritarian state power and socialized medicine will ensure that rapid and systematic elements are in place for the coming together of a biobank that combines genetic and clinical data and tissues from Malay, Indian, and Chinese patients by a target date of 2020." The collating of multiethnic databases and tissue samples, Yang claimed, "will help make Singapore a platform from which to introduce drugs into Southeast Asian markets." Instead of a standalone biobank (such as deCODE), Singapore is building an integrated biomedical ecosystem by mediating experiments that are proliferating in China and India (more than in Southeast Asia).

As noted above, China is very protective of its biological resources, and it does not permit the export of human samples, especially to the West. Here Singapore steps in as a research middleman who gains access to Chinese health data and is able to culturally manage PRC sensitivity such that the use of its health records should be of benefit to China and Chinese people. As mentioned earlier, PRC-born, Singapore-based scientists have easier access than most to Chinese health records, thus enriching Singapore's DNA and cancer databases. The information on genetic variants allows researchers to find biomarkers that they claim will ensure at least a 60 percent success rate for earlier phases of tests on novel customized drugs.

As a center for research on "Asian" cancers, Singapore has drawn contract research organizations (CROs), which handle outsourcing for clinical trials for drug corporations. Many trials that are run in Hong Kong and Taiwan focus on nose and throat cancers that disproportionately afflict Chinese populations. Given unreliable quality controls in PRC laboratories, over a hundred Chinese CROs have turned to Singapore to run experiments on new cancer drugs. Access to mainland Chinese DNA is central to the growth of clinical trials in Singapore. Scientists from Singapore help oversee ethical regulations of clinical trials in India. The intertwined scientific and economic strategies position the island as both a nexus and conduit for spreading best practices in clinical experiments in the region.

Williams claimed, "An Asian genetic architecture is much more valuable than biobanks in Euro-America because they do not carry Asian genetics." But as I have argued above, there is more to it than the furnishing of variegated DNA of the Asian dragon. Ethnic genetic collectivities are immutable scientific artifacts as well as bio-investments, while the ethnic design of the database builds domination through its ability to capture, produce, sum up, and prognosticate on DNA for a big swath of Asia. In short, the spread of computational biology, the competition of biobanks, and the demands of big pharma are all coproducers of this "plug and play" platform that furnishes ethnic-associated databases for speeding translational research from "bench to bedside," and generally for "making *more* of life" in Asia.[33]

To the researchers cited above, the "true value" of an Asian DNA infrastructure lies in its recognition of Asian peoples as worthy subjects of cutting-edge medicine. They were inspired in part by Venter, who in his guise as "the god of small things"[34] called for, as Williams recalled, "gathering up as much information as you can." He has been trawling the Pacific for microbes to reengineer into pharmaceutical products. Scientists in Asia want to beat him to the chase when it comes to producing data on Asian life forms that, having been gathered and calculated, are generative of diverse values beyond that of treating disease.

Conclusion

The Singapore and PRC scientists mentioned above deploy sequencing techniques to fold the reservoir of biological potentialities in Asia into an emergent, recombinatory, and mobile DNA database. The infrastructure deploys the ethnic heuristic in different registers. First, the network of ethnicity becomes a supple membrane coextensive with the network of genetic data points. Second, ethnicity is rendered an immutable mobile that circulates databases beyond tiny Singapore, making the infrastructure at once situated, flexible, and expansive. Third, the ethnic signifier carries affective value that enhances a sense of what's at stake in the building, mobilization, and implications of such Asian databases. In short, the origami-like folding together of multiple, flowable, and performative data points shapes a unified topological space of biomedical "Asia."

Meanwhile, American genomic science is not averse to the cataloging of DNA variability. Indeed, Yang's population genetics lab is one of four selected by the NIH to develop statistical research on DNA (the other participating centers are Cambridge University, Oxford University, and the University of Michigan). Compared to Eurocentric studies, Yang noted, the key contribution of his lab is the data on multiethnic associations for disease studies and drug reactions. Questioning the applicability of the DNA discovered for European populations to non-European populations, Yang claimed, "We are leaders in the game of trans-ethnic studies."[35]

Indeed, in positioning itself as an Asian biomedical hub, the Biopolis complex invests in the affects of biological difference and ethnic belonging.

Not surprisingly, the world of digitalized science rekindles abstract feelings about genetic exceptionalism. The spirit of the experiment also seems very old, relying on discourses that project an Asian genomic history of the body-genome-environment complex back in time among primordial "races" that were always in a state of flux.

This making and circulation of ethnogenomic identities raise anew the question "What is Asia?" By tracking the ways in which collectivities are defined and relations are conjured, revealed, re-formed, modeled, and predicted, Asian geneticists are shaping a novel concept of Asianness that is driven by scientific optimism. For Yang and his colleagues, terror incited by the wild things lurking in the 0.1 percent of the human genome can be managed by catching them in a novel web of corporeal and algorithmic self-readings.

TWO

AN ATLAS OF ASIAN DISEASES

A popular view is that the life sciences in the non-Western world (excepting Japan) suffer from poor scientific governance, a lack of regulation that contributes to unethical practices in the lab and in the field. At a transnational level, ethical debates about biocolonialism have protested against the theft of indigenous knowledge and resources.[1] When it comes to Asia, Francis Fukuyama has claimed that stem cell research has spread there as a form of "ethical arbitrage," though he averred that Singapore has a favorable regulatory climate.[2] The actual ethical configuration at the Biopolis complex is in practice a combination of global bioethical standards on the one hand, and the social ethics of national health care on the other.

When I first met Dr. Edison Liu in 2006, he flatly declared biomedical research in Singapore as less a force for profit making than it was a "canary in the coal mine." At that time, the most recent canary was the bird flu (and SARS [severe acute respiratory syndrome], still a fresh, searing memory), and the coal mine was and remains the Asian ecosystem, ground zero for all kinds of lethal diseases. Science is properly reminded of its role as a biosentinel for a whole region,[3] and of its need to develop interventions specific to it. For Liu, biomedical sciences are bound up in the technical elaboration of medical data into an early warning system for populations living in an Asia rife with potential afflictions. To develop science into a biosentinel infrastructure, in his estimation, is a matter of profound responsibility to Asian peoples, to be evaluated first for its ethical rather than its economic value. How much of this is a sales pitch, and are there really things that Biopolis can do that helps to maintain affordable health care for aging populations in Singapore and in Asia at large?

Biotechnology, especially the ability to reorganize bits of life, has changed the rules of the game, ethically and otherwise, pitting the powers of corporations against the sovereignty of nations to protect against biopiracy.[4] The role of the state in establishing control over biomedical research varies greatly across the globe, each site contending with the particular mix of scientific expertise and moral interests in its nation. As mentioned earlier, in Mexico, mestizo blood as a form of "genomic patrimony" is transformed into "mestizo genomics," symbolizing national unity.[5] In *French DNA*, Paul Rabinow notes the ethical value of blood and organs cast within a flexible notion of national patrimony.[6] In Singapore, however, the ethical reasoning is not about the sanctity of the genome, but about the potentiality of a modular and variegated DNA database for configuring a regional market while responding to the health needs of citizens. In other words, the fashioning of transethnic genomic information by Biopolis has dual goals of evolving the scientific enterprise to extend into other domains of "social trust."[7] Thus, Biopolis's quest to shape and control the life sciences has moved beyond economic entrepreneurialism, to become a bellwether of health threats to the domestic and regional contexts, a necessary role to the elaboration of science as a key substantiation of an ethical governmental enterprise.

And while clearly, despite Liu's rhetoric, the growth of the bioeconomy remains paramount, the Biopolis complex cannot be framed as exclusively in the service of biocapital or as a biosentinel in a biosecurity apparatus, but is also the complex negotiation of both roles in shaping a geo-biosocial project. Biocapitalism and social ethics achieve adjacency as apparatuses of mutual capture, each trying to capitalize and synthesize unevenly with the other. Indeed, the political justification for growing a bioeconomy in a competitive global arena is that research findings will address the silent explosion of terrifying diseases, which, among other things, threaten an improved way of life for billions of Asian bodies overlooked in Euro-American research. Importantly, I am not claiming an ontology or continuing an essentialism in Asian life sciences. Rather, I trace how bioscience practices in Singapore configure a localizable space stitched together by Asian ethnic identifications, relations, and perspectives.

Rather than a reversion to ethnoracial essentialisms, the name of "Asia," when attached to a catalog of serious diseases, acts as a global positioning device. The atlas of "Asian diseases" names an intervention into a Euro-American biomedical body that has been abstract and universal rather than historical and particular about human genetic variation. The novel biosentinel space deploys race/ethnic variables as coded elements of proximity in a new grid of material and ethical calculations, as well as a technique of emplacement that allows for appropriating an emerging geo-biosocial space. As Foucault's notion of "heterotopia" maintains, calculated spaces have different functions but are juxtaposed to other spaces (e.g., corporeal, ritual, archival) and times (regulated temporalities and flows).[8] Therefore, researchers aligned in this localizable biomedical space are not merely generic scientists working for the broader interest of a generalized humanity. Cosmopolitan science is realized through situated problematization, and Biopolis, with its atlas of Asian diseases, is effectuating a reproblematization and reassemblage of heterotopia of disease sciences.

This chapter looks at the different practices—a combination of international regulations and ethnic/Asian elements—that shape an emerging zone for naming and investigating "Asian" diseases. The domestic and international are not opposed here but rather name elements through which the Biopolis complex has taken its contingent form. Biopolis certainly would not have been possible without the adoption of international standards of bioethics for working with data, tissues, organs, and experimental subjects. Perhaps Singapore, more than any other Asian site, is best able to enforce regulatory regimes for governing reputable research and business, thereby creating a zone for the circulation of data, tissues, and experiments for a broader Asia.

A digital atlas of "Asian diseases" is both a strategic configuration of an emerging drug market and a symbolic mapping underpinned by, in Pierre Bourdieu's words, the "logic of reciprocal exchange."[9] An economy of Asian diseases unavoidably entangles market and social values in that while researchers hope to capitalize on multiethnic medicine, it also crystallizes the symbolism of movement, mutuality, and care across the region. From the vantage point of those on the island, a focus on "Chinese" diseases drives medical tourism, while reassuring the majority ethnic Chinese domestic population that its health needs are part of the mix. By enrolling the demographic heft and disease incidence associated with Chinese populations in China, researchers develop a robust database that attracts corporate investments in novel therapies, as well as clinical trials. But we should not misrecognize the symbolic capital of the disease economy that comes from linking Singaporean and People's Republic of China (PRC) genomic datasets to make an

"Asian" cancer catalog. The cancer cartography is a new symbolic economy that invests Asia as a site of many emergent values. By thus differentiating from a "universal first world body," Biopolis assembles the data, information, experiments, and meanings of "Asian" diseases, configuring a potential cancer research market as well as an emerging region of reciprocal research.

Bioethics in a Multiethnic Society

In order to convert all Singaporeans into an "experimental population,"[10] the government established a bioethical regime prior to the rise of Biopolis in 2003.

In the words of Dr. Williams, the expatriate American biostatistician, Singapore has "vigorous standards even more stringent than the FDA, follows Anglo-American guidelines on patients' rights, and fastidiously maintains medical records." In 2002, the Bioethics Advisory Committee (BAC) quickly established the tone of bioethics surrounding stem cell research in Singapore. Six of the U.S.-approved lines for research were developed by Dr. Arif Bongsu in Singapore. The government sought to capitalize on this by adopting British legislation for embryonic cell research, which included a ban on the use of human embryos beyond fourteen days of age as well as a prohibition on human cloning.[11] The BAC oversaw an extensive consultative process that surveyed dozens of religious and professional groups, exploring ethics of genetic manipulation in relation to notions of when human life begins. Focusing on four religious groups (assumed to largely overlap with ethnoracial categories), bioethics experts found that religious attitudes varied from beliefs that life begins at conception (Catholicism) to that it begins four months after birth (Islam). Only a small Catholic constituency, convinced that life begins at conception, and perhaps already primed by Catholic institutions about the new technique, expressed ethical objections. Religious leaders of the other three surveyed majority groups, which represented the vast majority of Singapore's population—Buddhism, Islam, and Hinduism—raised no objections to embryonic research (about which they were perhaps less informed). Under this nonobjection, they were assumed to be "for" stem cell research. The BAC framed its need to attain an ethical position that is "just" and "sustainable." By justice, the stress is on obligations to respect "the common good" and to share in the costs and benefits of this endeavor. By sustainability, the obligations are framed toward "the needs of generations yet unborn."[12]

This relatively brief window of debate was deemed to have sufficiently aired the ethical objections by major groups. By invoking John Rawls's "theory of justice,"[13] the BAC interpreted that what was being governed in the name of justice as fairness and with the telos of common good, was "life," its definition, point, and status, relative to living and health. The governance of life was no longer just a question of social benefits, tax codes, and income brackets, but a question of allocating a specific type of civic resource, that is, unwanted embryos needed for research. Such ethical discussions seem to have persuaded citizens already habituated to trust the state as always acting in their best interests to go along in a "pragmatic" sort of way, believing that the shift to biomedical sciences was inevitable for safeguarding their biological and national futures. The way was cleared for the state to proceed with stem cell research.

I interviewed a few vocal critics, as a robust bioethics debate never actually happened on the island. A Singapore academic reminded me that religious institutions have been excluded from public debate since the days of abortion legislation (in the 1960s). Whatever little resistance to the BAC there was came from Christian intellectuals who were probably more informed about bioethics from their overseas contacts. I spoke with Quan, a Catholic organizer who had debated Alan Colman on the radio. He told me Colman avoided his question: "How serious are we to explore the bioethics issues when events have already happened, to justify the decision? The ESC [embryonic stem cell] scientists claim the fact that 30 percent of fertilized eggs don't go to term as justification for their carrying out experiments." As a Catholic, Quan maintained that human life began at conception and that there was "no difference between therapeutic and actual cloning." He noted, "They thought they were sitting on a gold mine; the government felt that they can farm out rights to companies." A dissenting oncologist noted that the human embryo rule of fourteen days "seems arbitrary" in determining the development of the nervous system, and as a Protestant, she too believed that life begins at conception. She commented dryly that the BAC information was posted on the web, "but whether people felt their views were heard is another matter." There was worry about being "on a slippery" slope of an uncertain science experiment. A spokesman for the Catholic bishop observed that there was "little dissension as biotechnology is an economic good, and Singaporeans are very pragmatic."

Singapore's much-vaunted "pragmatism," a self-description of being tiny, under siege, and with a *kiasu* anxiety of "losing out," is the overall theme of the public going along with BAC guidelines. The majority ethnic Chinese were purportedly represented by the Confucian maxim that stresses the need "to sacrifice the individual life for the good of society, that the rights of society supersede the rights of the individual." The main issue was how truly pragmatic was this biomedical initiative? Quan asked, "Why use public money? If this is a viable venture, it is up to the business to persuade private venture capitalists to provide funding." Despite some skepticism, Singapore's cloning bill was a prelude to the fielding of genomic materials, thus creating a framework by developing licensing and regulation to reduce the possibilities of ethical problems. Additional legislation covers organ donations, a more uncertain achievement than bioethics.

Send in the Organs

Singapore's quest to build human tissue banks is a fraught enterprise that compels transborder collaborations for therapeutic research. In procuring organs and tissues, researchers are stymied because of the island's small population, low trauma rate, and ban on the sale of human organs. An earlier form of preemptive pragmatism was the 1987 Human Organ Transplant Act (HOTA), which introduced an opt-out system under which a person between the ages of twenty-one and sixty is presumed to have agreed to donate his or her kidneys, heart, liver, and corneas upon death, unless he or she has opted out. An academic noted, "It is quite pragmatic—these donations are made only to citizens." Muftis were persuaded to claim that organ harvesting was not against Islamic injunctions, but many Malay-Muslims reject the harvesting of organs, preferring to be buried with their bodies intact. An earlier policy was that "communities that do not donate have low priority as recipients." In 2008, an amended HOTA sought to get more Muslims to opt in to donate their organs and give them equal access to transplant procedures. The "opt-out" system hoped to make organ donation somewhat routine but there has been substantial resistance among citizens of all ethnicities. Citizens expressed the right of loved ones, not medical experts, to decide to donate organs of the deceased as they deemed fit. "There is a minefield of ethics," Dr. Loh, an oncologist, remarked, because of additional complicating factors such as the fact that, in a situation of low numbers of cadavers and high demands for kidney transplants (among mainly Chinese patients), minority Malay-Muslims are an important pool of potential donors, as they tend to have the statistically highest rate of accidental deaths (from motorcycle accidents).

The focus has therefore shifted to living donors who are related to potential recipients of kidney or liver transplants. The demand for kidneys is especially acute because thousands of patients on the island, mainly ethnic Chinese, suffer from kidney failure each year. Initially the focus was on living donors who have an emotional link to recipients, thus tending to relatives, friends, or even coworkers. A transplant ethics internal review board (IRB) seeks to weed out any possible monetary inducement, but hospital officials admitted that when kidney recipients and living donors come from overseas, they may not have the means to check on a nonrelated donor who has been paid off by a wealthy patient. Loh continued: "We may be opening up a can of worms, with the possibility of the rich exploiting the poor." In order to expand the pool of potential living donors, in 2009, HOTA was further amended to include compensation to those who donate their kidneys to patients they do not know. The "donor welfare scheme" was intended to funnel much-needed organs from hospitals in the region so that Singapore could become, according to Dr. Chua, an official of Singapore General Hospital, "a medical Mecca for patients from Russia, the Middle East, and Asian nations."

Assembling human tissues for research versus for therapeutic purposes may be just as challenging. The proliferation of surveys in hospitals, universities, and research labs has increased the number of people willing to be tested. But building a tissue bank for research is an ongoing and uncertain enterprise. Other than women routinely permitting the donation of umbilical cord blood or "hospital waste," even the donation of blood products to science may be rather uneven.[14] Among different ethnic groups, blood is considered a precious family resource imbued with cultural significance that should not be easily made available for nonkinship purposes. Doctors report that unless they are patients with fatal diseases, most citizens will not give freely of their blood. Furthermore, some physicians themselves may have ethical reservations about handing over samples that may end up as research materials for foreign drug companies.

Here, once again, ethnic discourse enters into efforts to induce more willingness to give blood and organs. Soft authoritarianism has not been able to corral human organs on the scale that is desperately needed, despite mandatory medical insurance, an opt-out rather than opt-in system, and donor compensations. Speaking with many researchers, I had the sense that, through discourses on ethnic-disease links, they hoped to prime the pump that will yield more contributions of human tissues. For instance, Chua, a public health expert, has done a study linking "genetic races with diabetes" by using ethnic categories derived from the national record number and self-identification to identify subjects who consent to blood studies. Chua's report "found links between Indian ethnicity and diabetes, obesity, and hypertension, similar to Indians in India and in the U.S." Furthermore, he added, "Similar genetic trends have been reported in studies done in Malaysia and China." The ethnic typing of diseases through mining existing database, demographic, and tissue records sought to leverage ethnic solidarity into a duty or desire to participate in state biomedical initiatives.

There is, however, the complicating factor that, for many Asian groups, kinship is materialized by blood with/from ancestors, an understanding of kinship as a reciprocal system that is different than the scientific remodeling of kinship symbolized by ethnicized genomes. So for ethnic groups, ancestry and kinship are two sides of a coin of heredity. This braided notion of heredity has been a stumbling block for the organ donation program because physicians emphasize the ethnicity of the disease (linked to the need for transplants). Given that the major barrier in organ donation is "kinship," in that people aim to keep blood and body parts safe from nonkinship uses, Chua suggested, perhaps an alternate recruitment strategy would be to mobilize ethnicity against kinship for the contribution of blood and tissues as resources for a larger collectivity. There is thus a kind of sundering and then resuturing, as it were, of kinship in order to finesse the system of organ donations.

Here science harnesses Benedict Anderson's canonical "imagined communities"[15] in the service of collecting organs. The postcolonial country has used blood, bloodlines, and ancestry variably to make kinship something like an ethnicized nation. There was an appropriation and reorganization of relations between multiple forms of kinship in reconfiguration under the signs of anti-imperialism, citizenship, and nationalism. But the twenty-first-century emergent forms of corralling national(istic) idioms for reconfiguring the scale and register upon which kinship is established and operative are at a regional collectivity.[16] Ethnicity is delinked from its colonial moorings and resutured in its relationship with ancestry, genetic predisposition to hereditary disease, civic duty, social status, and health in an emergent region.

In multiethnic Singapore, a strategy of appealing to the sense of tribalism can build on already-existing charity and philanthropic activities orchestrated by ethnicity. A Red Cross leader remarked that, in his nation, the government recognizes that it is "not a melting pot. There is practical recognition of the world as it is, of people giving to their own kind." Here, the mobilization of the ethnic heuristic is also a strategy of engendering a collective relation to a body that exceeds that of the family. This is the "ethnicity" that is rendered vulnerable to afflictions of different kinds. This is a relationship of substance, in experimental and affective terms, rather than nation per se, in that it references a biological conception of the continuum of relation that does not directly reference political claims about identity. Rather, it reconceptualizes the genetics of Asian ethnic diseases at a crucial juncture: a site in which Euro-American genomic entities are imagined as poised to capitalize and exploit Asian genomic information.

The National Tissue Bank obtains samples with consent from public hospitals and clinics and delivers them to institutions or programs that bank the material. The role of the medical researcher is to cull records for the necessary data that accompany the sample, as to well as to make the ethnic association with given diseases. Probably, to construct an ethnic profile, the researcher uses a mix of the medical records and personal ID information for details on gender, age, and ethnicity of the male parent.

Access to and uses of human samples are also highly regulated. Tissue, sera, and DNA samples are collected from hospitals and health centers, with donors filling out informed consent forms. The use of surplus embryos from fertility clinics, including aborted fetuses, requires the consent of parents who are not recompensed. Human reproductive cloning is banned, and therapeutic cloning (stem cell research) is allowed on embryos up to fourteen days old, following British bioethical guidelines. Hospital bioethics committees make regular checks of clinic practice and provision is made for conscientious objection. There is no black market in human organs in Singapore, and with the strict controls in place, Singapore is considered a respectable place to conduct stem cell research in accordance with international bioethical standards.

Private companies can secure genetic materials by approaching the National Tissue Bank with research proposals and collaborating with local institutions for clinical trials. For instance, applications to the national cancer center need to involve a local collaborator in a research initiative. Intellectual property (IP) claims by banks are made on a case-by-case basis, and, according to the head of the cancer center, all IP claims are carefully scrutinized. The common practice is to make a claim as a "return to the community" in the form of share of profits and licensing fees. In short, from the bioethics protocol, to the opt-out system and the National Tissue Bank, to regulations controlling access to data and samples and banning the sale of tissues, the Singapore government has coordinated and streamlined different aspects to ensure an aboveboard process for using anonymized but ethnically specified biomedical data and materials.

Testing Humans and Animals

Bioethical guidelines for drug testing are also very stringent. Subjects volunteering for clinical trials must sign informed consent forms and cannot be induced with payments. Test subjects are paid for missed work and travel to the clinic. The strict guidelines mean that physicians contracting to do clinical tests for drug companies sometimes cannot find enough subjects.

Besides the fact that the island has a population of only 5.5 million, Singaporeans who are not sick are generally not interested in being test subjects. A doctor notes that "Singaporeans are educated but not sophisticated; the word 'trial' makes them think of being guinea pigs." Here Singaporeans seem to act like Americans, whose avoidance of being "lab rats" have pushed many clinical trials overseas, to places such as Singapore where patients may be slightly more amenable to research apparatuses. Nevertheless, Singapore is a trusted site for clinical trials, especially in the initial phase, because of excellent facilities, record keeping, and follow-up with patients. Many drug companies now spread clinical trials throughout Asia, using India or China as the site for Phase III testing, with a Singaporean clinic as the coordinator or leading partner in these procedures.

Singapore has one of the largest animal facilities in the world located on an island. It is regulated by a licensing system that includes strict norms for animal experimentation. The inhumane treatment of mice would result in fines and being jailed, I was told. Ethics surrounding research animals do not have to deal with the animal rights movement, which does not exist in Singapore. Some British scientists welcome the prospect of working in Singapore without threat of disruption by animal liberators who are considered, under Singapore's stringent laws, to be "terrorists." "Asian societies tend to be more pragmatic; we feel it's more important to find the causes of diseases than to worry about the rights of mice," said a local representative of Johns Hopkins in Singapore.

At a 2005 international meeting on clinical and lab genomic standards, Singapore was judged as participating in the process of "global harmonization" of standard controls and best-practice guidelines so that patients, clinicians, and researchers can extract the maximum benefit.[17] Enforcing best-practice guidelines in the biomedical world, along with "pragmatism," "track record," and "keen foresight," it was claimed, has made international businesses entrust their capital and invest their future in the city-state. Indeed, global recognition of Singapore as having the best record for bioethics and IP protection in Asia authorizes the island as a zone for biomedical science in Asia that also hopes to safeguard political interests.

"Gene War": The Right to Know

Perhaps politics, more than IP, drives the rush to develop Asian genomics's capacity to prevent breaches of biosovereignty.[18] The following interview with Dr. Lee, a population geneticist trained in the United States, reveals anxieties heavily tinged by what appears to be a kind of neo-eugenic fear pointed against "weak" Asian constitutions. She noted that the Human Genome Organization was tracking human DNA resources worldwide, but Asian scientists should control genetic knowledge in and of Asia. In her view, we are in a "new era now as social and economic aspects previously ignored have come into play. . . . There are also legal implications of our capacity to predict illnesses that may not yet be visible." She thus posed political questions about who should or would own genetic knowledge of Asian peoples and thus come to control the customized medicine that develops from it. The main issue for her was the configuration of biomedical knowledge and how it serves to potentially exacerbate existing geopolitical differences in a situation where "ethnic" differences are deep, bandaged-over political wounds.

She mentioned a controversy surrounding international collaborations on genetic research in China in the late 1990s. In 1997, a Chinese doctor from Anhui Medical University (without consent from Beijing) shipped thousands of human samples to U.S. companies. This unauthorized sale ignited a firestorm in China's biomedical community, with some calling it the "gene war of the century." A group of U.S.-based Chinese geneticists accused the Chinese media of stirring up a DNA hysteria and that the report of foreign access to Chinese DNA was "replete with factual error, misrepresentation, and overstatement." The Chinese protests express faulty assumptions that China has the richest genetic resources in the world and that gene identification is the driving force in drug discovery. Furthermore, the geneticists argue, there are severe limits in Chinese science: poor documentation as well as limited population genetics and genetic epidemiology in China. The science authorities have displayed poor capacity to support and guide independent pharmaceutical research and development. Although the Chinese media invoked the "gene war," some Chinese scientists expressed concern that foreign involvement could lead to "misuse" of Chinese DNA, calling for a ban on exporting samples.[19] The domestic furor mirrored fears of being impinged on once again by Western powers, mirroring the semicolonialism in China's ongoing "century of humiliation" at the hands of foreigners. In 1999, a new law banned the export of Chinese DNA for research outside China.

Two decades later, despite the rejection of the veracity of the accusation by PRC geneticists, the story is still salient in and beyond China. Lee noted that some Chinese people thought that foreign access to "indigenous samples engendered fears that the groups may be portrayed as genetically weak." Such concerns about "external" exploitation of Asian DNA underscored her insistence that scientists in Asia should control the genetic resources in their geographic area. These genetic resources find expression in phylogenetic suites of traits, including ethnic-specified disease-risk profiles.

Lee phrased it this way: "Who has the right to know? The Asian context is different from the West: we do consultations about such issues." Authorized experts have to oversee the governance of such potentially explosive information about "groups" (self-identified race or continental ancestry) and their gene identification that is the driving force in drug discovery. Against the feeling of a Western master ordering of human genetic difference, she argued, Asian scientists mapping and delimiting Asian ethnicities and samples would resist a racist and exploitative misrepresentation by their Euro-American counterparts. Central to this is an elaboration of Asian scientific institutions— and by extension, the Asian states that fund them—as the proper interpreters and spokespersons for Asian genetic riches and vulnerabilities.

At the Genome Institute, Dr. Wu (who bar-coded ethnic DNA) added the political and economic dimensions of controlling genomic patrimony. "Most countries see themselves as owners of their own national genomes. China, for instance, is beginning to place limits on tissue samples acquired by outsiders through collaborations. They realize that disease populations are of great value; there is the element of economic value and output now attached to them." One way for Biopolis to secure and capitalize on national genomes is to enroll researchers in the PRC with whom they cultivate understandings of shared "Asian" interests in attending to the medical traits and needs of Chinese people at home and abroad. Given the unfortunate history of the Malaysia-Singapore rift in 1965, a similar transnational collaboration on Malay genomic research has not (yet) been feasible. By deploying the ethnic heuristic, raw materials can be recruited, hopefully, through an attempt to resituate data and samples in a collective substance, vulnerability, and hope that transcends national borders, rather than in the confines of the family or the island.

Asian Genes and Asian Diseases

Biopolis seeks to create disease-driven maps of populations in Asia, as part of its competition with rivals such as Bumrungrad International Hospital, in Bangkok, to dominate the booming Asian medical tourist market.[20] At the Genomic Institute, a spokesman said that researchers were driven by "the scientific assumption that Asians and Caucasians are distinctly different. We develop tailor-made medicines for Asians."

This approach was confirmed by a leading American scientist, Dr. Thompkins at the institute, who explained that researchers gather samples and do the follow-up work on "genotyping, collaborate between genotyping and building an ethnic genomic map in a cohesive and real-time fashion." The institute focuses on cancer (Liu's specialty), that is, breast, liver, and gastric cancers that are prevalent. Not only were the multiethnic populations critical for discovering genetic variants, there was "a wealth of samples in hospitals for collaborative efforts with foreign companies." In the early years, companies such as Eli Lilly, GlaxoSmithKline, Schering Plough, and Pfizer mainly had marketing offices but not wet labs in Singapore; but a decade later, as a result of the ethnic genomic map, forty corporate labs were linked to research projects at Biopolis.

Researchers in the United States and in Asia have pointed out that Asian populations display markedly different reactions to drugs than Westerners. Access to multiethnic populations, they claim, is thus critical: "Ethnicity is a very important element because some genes are expressed more strongly in some populations than in others." Doctors point to the prevalence of fibrosis and meningitis among whites, and the diseases' relative absence in Singapore. Lee mentioned that Western patterns of cancer (e.g., prostate cancer), heart disease, and other common diseases have different profiles than in Asia. Infectious diseases too may be uniquely expressed by different ethnicities. A spokesman for Johns Hopkins Singapore, which opened in 2000, noted that the Epstein-Barr virus in Caucasians causes mononucleosis, but in Southern Chinese it causes nasopharyngeal carcinoma (NPC), or nose and throat cancer. "The mechanism is still unknown but we are developing some experimental therapies. NPC rates are as high as twenty to thirty individuals out of one hundred thousand in the Singapore population. We study only the sample that has NPC." He explained the use of the ethnic heuristic: "We can't fix SNP [single nucleotide polymorphism] variability on individuals, but must infer from a category." Knowing the genetic variation (metabolic process) of an ethnic group will allow doctors to anticipate a drug's effect or to identify people at risk. Thompkins stated that drugs have a genetic component and therefore need to be modified for dosage, especially for "Asian" diseases such as hepatitis B. "Asian populations are a pull for drug trials to be conducted here. Drug companies seek different markets and needs. There is the understanding that Asians are in a different localization." He summed up by saying that pharmacogenetics can develop different classes or dosages of drugs for Caucasians and Asians.

As defined by life scientists at Biopolis and the distribution as calibrated by population geneticists and biostatisticians, and as visualized and generated by lab technicians, the ethnic heuristic has been used to diagrammatically plot correlations between groups and diseases for whatever experiments research-ers are pursuing. Thus, physicians in Singapore can and often do reel off fright-ening figures of major diseases that disproportionately affect people in Asia. The continent, they pointed out, has the world's highest incidences of stomach cancer, lung cancer, hepatitis B, and infectious diseases. Ethnic Indians (mainly Tamils and Sikhs in Singapore) are considered especially vulnerable to heart diseases, and Muslim-Malays to obesity and diabetes. Chinese, Indians, and Malays are considered genetically susceptible to hepatitis B and hepatitis-linked cancer. "Hepatitis B is an Asian disease. Seventy percent of worldwide occurrence is in Asia, and the most prevalent rates are in South China," where a huge percentage of people under forty years of age are infected. A spectrum of cancers—gastric, liver, and esophageal—is found to have the highest fatality rates among mainland Chinese from Guangdong and Fujian Provinces.

Singaporean researchers mobilize around such "Asian" diseases by con-necting specific ethnicities to genetics, metabolics, immunology, and geogra-phy. The general goal in labs is to pinpoint "unique ethnic genotypes" in order to strengthen defenses against "Asian" diseases. An international figure in the life sciences lends his presence to the strategic viability of ethnic-identified gene therapy. Alan Colman, the Scottish doctor who helped clone Dolly the Sheep, is working on cell-replacement therapies for insulin-dependent diabe-tes and cardiovascular diseases, both of which are identified as diseases men-acing Asian ethnic populations. The fields of genetics and immunology are converging as well in the quest to develop new diagnostic and therapeutic methods.

New models of Asian cancer biology trace not only mutations but also the immune systems of Asian patients. Dr. Tan, a Harvard-trained stem cell ex-pert, developed a therapy that enlists the patient's own immune system to fight cancer tumors. In an interview at his lab, he argued that family and eth-nic genetic compatibility has a significant role to play in both protecting and reducing side effects in cancer treatment. In throat cancer, a "mini-transplant" with stem cells from a sibling donor can boost the patient's immune system during treatment by killing some cancerous cells or reducing patient exposure to radiotherapy or chemotherapy. The same logic of ethnic or family stem cells protecting against rejection, he remarked, underlies the use of a relative's bone marrow for treating leukemia patients. Thus, techniques that connect ethnicity to genotypes and immune systems seem a promising strategy for the experimental treatment of Asian cancers. Such ethnic indicators, researchers believe, also generate potential values by narrowing the samples that will ensure successful testing for particular drugs and thus further enhance the attractiveness of Singapore as a center of treatment for major Asian diseases.

Perhaps my own Asian ancestry drew such candid and unapologetic claims about "Asian" differences in disease incidence, disease form, and need for customized medicine. My impression, however, was that the researchers genuinely believed they were on to something, that they were mining a special vein previously ignored in biomedical research, and that Asian diseases constitute a new area for novel pharmaceutical development. The message of Asian diseases was also part of the promise that Singapore's medical tourism sector was developing niche, high-end therapies customized to sick bodies in Asia, especially in China.

Cancer Taxonomy and Scalability

A revolution in cancer research turns on decoding genetic mutations and understanding molecular changes that propel rogue cell growth.[21] Cancer is increasingly viewed as not one but thousands of specific ailments, each with its distinct molecular signature. Genomics and computational biology have come together to map a new Cancer Genome Atlas. In this new landscape of cancer, according to the U.S. physician Siddhartha Mukherjee, "The entire genome of several tumor types, every single mutated gene[,] will be identified."[22] Researchers in Singapore are finding ways to categorize tumors according to inherited mutations and somatic mutations, thus providing a roadmap as well to the region's collection of cancer forms.

In many ways, Singaporean biomedical science is an adjunct to supportive mechanisms for U.S. research. An American oncologist at the Johns Hopkins Cancer Center in Singapore was quoted as saying, "Singapore is exactly the kind of place where a scheme like this would work. . . . A national database, along with a detailed chart of genetic mutations among cancer sufferers, could only serve as a hugely effective new weapon in fighting cancer."[23] Institutes, hospitals, and clinics have come together to hunt for genes and discover biomarkers for "Asian-specific forms of cancer" (breast, stomach, throat, and lung, among others). U.S.-Singapore collaborations are configuring a new frontier of cancer research, one that is not shy of invoking ethnic risk profiles and therapies.

In the new cancer, gene defects gain a pronounced materiality as cancer samples and data that interact with prevailing ethnicity. "Ethnicity" operates experimentally as a composite term variously invoked to locate, signify, and relate particular genetic mutations, epigenetic effects, molecular signatures, and geographic zones. Oncologists and researchers in Singapore repeatedly claim that "Asian-specific forms of cancer"—breast, throat, lung, liver, and stomach—were most pervasive among peoples identifiable as "Chinese." For instance, nasopharyngeal carcinoma has "an unusual degree of geographical localization that has earned it an unfortunate nickname, 'the Canton tumor,'" in reference to the region of South China where these cancers seem to concentrate. An oncologist at Biopolis explains that NPC "exhibits unusually high prevalence in southern Chinese descendants—mostly in Southern China and Southeast Asia—with an occurrence rate about twenty-five times higher than that in most regions of the world."[24] Some health workers suspect its link to a greater susceptibility of ethnic Chinese to SARS. Another "Asia-Pacific" disease is liver cancer because of the prevalence of chronic hepatitis B and C in Asian populations, and 80 percent of liver cancer worldwide occurs in the region, especially among people originating from Southern China. Thus is drawn a map of Asian cancers as distinctive, brimming with threat and opportunity, crying out for research.

Cancer genetics are so variable that the satisfaction of statistical significance of a particular mutation depends on compiling data from hundreds of thousands of patients. Given the demographic weight associated with "Chinese" populations, the coding of "Chinese" cancers makes it flexible or ambiguous enough to scale up or down, as research demands. This slippage between Chinese and Asian is shaped according to the abstractness of scientific truth claims that facilitate flexible deployment of biomedical science and justification across many sites and forms of cancer.

Cancer, of course, has a fatal presence in the health imaginaries of ethnic Chinese because it is a dominant presence in the region. The ethnic coding of cancer science therefore links, in abstract and material ways, researchers in Singapore, China, and Hong Kong for whom the cancer databases animate affects of being "Chinese," at risk, and at work to remediate the "Asian" cancer problem. The cultural preconception that there is a powerful connection between being Chinese and being susceptible to particular forms of cancer offers Chinese people a chance to be increasingly in charge of their own health problems and for ethnic Chinese scientists *beyond* China to undertake new battles against cancer risks. In an atmosphere of mutual receptivity and benefit in shaping the field, researchers have begun to cooperate by documenting, tracking, and studying so-called Chinese cancers.

The Chinese Cancer Cartography

Here I present a case of cross-border collaboration between oncologists at Biopolis and mainland China to map various kinds of cancers prevalent among populations identified and identifiable as "Chinese" or "Han." "Han" refers to the majority "nationality" in the PRC, a country that is composed of fifty-five official nationalities, determined by the Chinese state (see chapter 9). I will also highlight the techniques of scaling and blending of categories that are authorized methodologically and interpretively. It bears noting again that the heuristic and pragmatic value of ethnic DNA coding in territorializing a cartography of cancer in no way for scientists means a causal or primordial link between population and disease. Rather, it is a way of correlating differential risks and incidences of ailments with "ethnic" populations widely scattered. The reasoning is that, due to a high incidence of rather specific forms of cancer, the huge populations, and geographic location that are identifiable as "Chinese," the ethnic heuristic seems both unavoidable and pragmatic in cancer genomics.

Indeed, when China is drawn into the cancer research assemblage institutionally, the potential population from which medical samples, data, and experimental subjects can be gathered seems limitless, and the "Chinese" sample set in Biopolis balloons far beyond the patient records of ethnic Chinese citizens in Singapore to potentially absorb "Chinese" bodies elsewhere. While scientists also study cancer manifestations in different populations, the Biopolis research focus on Chinese cancers is at the core of the emerging field of trans-Asian oncology, as it draws popular political support and is buttressed logistically by transborder collaborative efforts that intersect in the Biopolis complex as a coordinating node.

Dr. Lin, a China-born, leading cancer researcher at Biopolis, told me that Singapore is in a unique position when it comes to Asian cancer research. Although, say, Hong Kong has the same kind of resources, and is the much vaunted "gateway" to China, Singapore's significance is its approach and what it is banking on. Singapore's government has identified a genomic approach to cancer as a strategic area of research, with its interlocking institutional, heuristic, governance, and partnership facets. This paradigm, the data it will yield, the applications for that data, and even the market strategies prior to therapy or device development are missing in Hong Kong. The Biopolis strategy may not be foolproof and is fraught with risks of (commercial) failure, but yields in the research sector will ramify across so many other domains of the island and Asian life.

Lin led cancer research on populations—Indians, Malays, aboriginal groups—but his main focus was on compiling a huge database on cancers in Chinese populations. In mapping cancer genomics, the deployment of the ethnic heuristic seems to engender an affective notion of unified Chineseness in the far-flung, collaborative scope of his work.[25] Here we see how the research and sample gathering perform and foreshadow the claims of the project to speak to a widely conceived and geographically discontinuous and ungrounded vision of "China" carried in vulnerable genomes.

As noted earlier, Lin's data on Chinese cancers draw on samples from Singapore and China. The database is designated "Chinese" not due to a belief that there are no biological and cultural differences between samples, but because the repetition of a Chinese code formalizes a new terrain of cancer DNA research, setting it off clearly from non-Chinese cancer data. His collaborators include researchers from across China: Fudan University in Shanghai working on lung cancer; researchers from Anhui Province studying psoriasis; and those from Zhongshan University in Guangdong Province focusing on carcinoma. His immediate goal was to have his lab be a major source of PRC genetic mutations and samples for a range of major diseases that seem to disproportionately menace peoples of "Han Chinese" ancestries. While the Han ethnic group is only one of fifty-five nationalities in the PRC, Lin's "Han Chinese" coded genome allows it to stretch across space and beyond the territorial borders of China, so that it is not geographically congruent with the territorial stretch of China, which includes many non-Han populations. Rather, it stretches out over a novel space defined in the movement of embodied Chinese genomes. Tracing histories of migration becomes a method for territorializing a genomic cartography. Beyond genetic defects associated with cancers, data and samples from the mainland will permit intensive work on the genetic bases of disorders such as Parkinson's, stroke, and schizophrenia among Han Chinese populations. It appeared that PRC and Singaporean researchers were collaborating on a growing list of serious diseases affecting ethnic Chinese peoples.

While China is a major source of samples for Biopolis-based oncologists, those in China sometimes worried that Chinese cancer DNA was an overly homogenizing code. Lin himself had published a paper that reports genetic differences between Northern and Southern Chinese (Han), the often-cited two major (sub)ethnic categories that crop up in the new genomics. He noted that the mitochondrial DNA of Han Chinese increases in diversity from north to south but that there is genetic coherence in all Han-identified populations.

Dr. Yu is one of Lin's lab researchers and also is originally from the PRC. She was still worried about the research practice in China of ethnic labeling at different stages of sampling. Trained as a bioinformaticist, Yu prided herself as especially skillful in mining data (SNPs, risk alleles) "to get novel information" out of samples. She nevertheless confided to me that "the situation is becoming quite tricky in naming samples as 'Han Chinese.'" She remarked that, "in China, samples are identified by the geographical areas from which a patient comes or in which he resides. With mass movement and migration, this is becoming a problem. But usually Central, North, and South China are areas with assumed homogenous samples" by researchers in the different sites. It remains uncertain whether such assumptions are actually to a certain extent borne out in data about contemporary migration.

What, I asked, was the research method for identifying a donor's ethnicity? Yu observed that, in her homeland, there was a social tendency to conflate residence and ethnicity or to reterritorialize ethnicity with territory. Furthermore, the range of ethnicities commonly invoked tended to be limited to "Han" and "minority" (*shaoshu minzu*), mainly Hui or Miao. "The doctors usually ask simple questions of patients, where do you live, or where are you from? Doctors tend to ask for self-identification (residential, hometown, or ethnicity based) but don't ask about parents' ethnicity. Of course, patients collected in big cities such as Beijing or Shanghai may be from elsewhere, as many come from different places and ethnicities. Even Han people have great diversity. But samples collected in Shandong or Henan (which tends to be a source of emigrants rather than a destination for migrants) were mainly from natives of these provinces. And ethnic minorities are quite obvious because of the geographical location of the sample collection, for example, the Zhuang from Guangxi, Miao from Yunnan." At different steps of sample collection and analysis, PRC researchers tended to inscribe a unifying Han identity onto diverse genetic samples, and when these were gathered together in Singapore, the PRC data are integrated with other pools of genetic markers drawn from ethnic Chinese populations scattered throughout Southeast Asia.

Lin is thus assembling a database on "Chinese" cancer mutations that is far-reaching as a master Asian DNA code and purports to represent a range of ethnic groups. Cancer maps of genetic defects not only promote Chinese ethnicity as an expanding biomedical category, but also configure a research field that uses situated notions of ethnicity (Chinese, Indian, Malay, etc.) as points of reference for tracking genetic variations and rates of susceptibility to different types of cancer. That is, various administrative understandings of Chineseness in different national political cosmologies are consequential in the research. Obviously, these ethnic terms themselves have colonial and nationalist underpinnings as well, and so they become reified and even expansive through being deployed as scientific collectivities.

Again, researchers believe not that cancer expressions come in ethnically defined categories, but that, as mentioned above, they are triggered by the interaction of genetic, environmental, and lifestyle factors. But for practical reasons and research, clinical, and administrative purposes, ethnic coded information on genetic defects is one tool among many for the development of customized medicine (see below). Computational genetics articulates other fields, such as molecular biology and stem cell research. Genomics is crucial to translation research, from bench to bedside, in vitro and in vivo, building chains of genetic causation and potential cures. In the clinic, genetic data can facilitate the molecular investigation of particular expressions of diseases in individual patients. Ethnic-coded bioinformation allows oncologists to check for drug trials that have already been performed on certain groups and to also better select the drugs that may work for different molecular subsets of cancer.

The search for genetic cancer flaws thus makes visible ethnicity as a biomedically risked quality, but at the same time it holds out the promise of developing customized therapies. Indeed, researchers talk freely about the so-called South China founder effect to account for the forms of cancer encountered in Asian patients. Key biomedical centers are located in Chinese-dominated sites— Singapore, Hong Kong, China—that have access to appropriate population samples and experimental subjects for research. Thus, when scientists in Singapore or in China persuade donors to donate samples for studies on Chinese cancers, they create an abstract affect of shared DNA, compelling feelings for donating samples needed for biomedical intervention.

The sharing of genetic materials from multiple sites variously invoked as "Chinese" cannot but stir professional investment and solidarity in cancer genomics. Scientists such as Lin and Yu are self-identified Chinese researchers investing their careers in mapping ethnic-identified genetic mutations for cancers that menace co-ethnic populations. There is an imaginary of shared genomic origins and social trust between co-ethnics designing a novel disease field under their professional and political control. Furthermore, collaborations with researchers in Singapore are a distinct career advantage for PRC oncologists, because as joint researchers they get published in international journals. Singapore-PRC collaborations have become critical for mainland Chinese scientists to gain international recognition for their work, because Singaporean researchers are among the most productive of the top five countries in terms of papers per thousand researchers.

The framing of oncology through an ethnic lens highlights ongoing tensions between researchers working at the biostatistical and the molecular levels, or between collective and individual data. Despite reservations especially on the part of classically trained biologists, both approaches, I was told, work in a synergy that is potentially productive in developing new diagnoses and therapies for Asian cancers. We see such tensions played out at the Duke-NUS Graduate Medical School in Singapore.

During the first of my many visits, the reception area of the Cancer and Stem Cell Biology (CSCB) area at Duke-NUS was decorated with posters that celebrate the opening of the center in 2009. One of the banners proclaims, "Our objective is to be a center of excellence in CSCB, basic and clinical translation studies, with emphasis on molecular interventions and cancers that affect Asians." I asked Dr. Fisher, an American oncologist at the medical school, how molecular research would be different for "Asians." As a self-described "hypothesis-driven international scientist," he explained that, "in cancer research, we are just scratching the surface; there are about one hundred cancer tumor genomes." He continued: "Researchers on Asian diseases, such as lung cancer, gastric cancer, carcinoma (liver cancer) that are common here, want to have access to patients." His goal was to get close to these samples and data drawn from "Asian populations" so that he can help develop the vaccines. While he scoffed at macro-level predictive data that Liu and Yin assembled as a form of "bird-watching, where researchers throw everything at the technology and see if it sticks," he does concede that the data have been useful in providing signals as to where to look. Indeed, the genome-wide subgroups help him zero in on the kinds of molecular mutations associated with Asian groups susceptible to particular forms of cancer, and thus he can refine steps toward personalized treatment.

From their different angles on genomic research, Drs. Tan, Lin, Yu, and Fisher shape the cancer map as "Asian/Chinese" in multiple ways: as an assemblage of cancer data, as patients, and as the public at large. "Asian" suggests specific DNA and expressions of subtypes of cancer as well as large amorphous collectivities in a global region. "Asian" also configures an innovative frontier in cancer research that, while driven by biocapitalist interests, cannot be disentangled from the politics of apprehension, optimism, and investment in a disease that threatens potentially billions of people.

Conclusion: For the "Asian" World

Donna Haraway has argued that scientific practices make visible the alignment of the technical and the political, as well as the partiality of location "in the sense of being *for* some worlds, and not others."[26] The spread of genomic technologies allows emerging Asian countries—covering some three to six billion of the earth's inhabitants—to address medical and environmental problems specific to the region. Genomic technologies increase the capacity of developing countries to secure the data warehouse of their populations and to allow sovereign nations to gain control of their genetic information and its interpretation. A major step is the charting and naming of lethal diseases and the development of treatment for people whose susceptibility to disease is conditioned by descent/membership in particular groups or from having evolved in particular environments long neglected by cutting-edge science. By developing a disease knowledge dedicated to the future well-being of a region, a specific conception of "Asia" or "Asians" as a site that is brimming with debility is turned (through biopolitical state intervention) into a new frontier for commercial life sciences.

In the process, researchers in Singapore challenge a first-world conception of the body that is taken as a given in biomedical research. Rather than universal, this body is specific and localizable in a particular space of knowledge formation. Rather than individual, this body is generated through determinations of ancestry, ethnicity, and descent that bind its fate with others. But the ethnic mapping of diseases is made universalizable only because Singapore has made itself a biomedical platform, with the implementation of a series of legal and bioethical regimes so that reputable research can be conducted and patents potentially developed for "Asian" cancer treatments. As well, through research collaborations with China, Singapore gains control of the circulation of critical data and samples from patients who represent a huge majority in the region.

While this pluralization of cosmopolitan science has been driven mainly by Biopolis's aim to extend its bioeconomy, there is also hope that identifying "Asian" diseases grounds life in Asia as worth living, and disease science as worth doing.

In the late 1970s, my older sister, then in her early thirties, found a lump blooming under her right armpit. She rushed uptown to have a biopsy at the Memorial Sloan Kettering Cancer Center in New York City. When the doctor announced that the cells were malignant, I fainted. She did not. She soon began the punishing schedule of surgery, radiation, and chemotherapy.

A few years later, my older brother, himself a physician in Singapore, was diagnosed with leukemia. He was soon transferred to the Fred Hutchinson Cancer Research Hospital in Seattle for a bone marrow transplant. I remember the frantic calls for blood tests to be performed by immediate family members scattered across the world. The four eligible donors included our aging father and three brothers. Bone marrow transplantation is a delicate procedure drawn out over many months. The chosen donor brother moved from Princeton to Seattle to be available for the almost daily siphoning of healthy blood cells through a long needle inserted into his hip bone. After a few months, the leukemia was brought under control, and the brothers returned to their respective jobs as surgeon and as professor. My sister had early onset breast cancer, and my brother had leukemia that required DNA matches from siblings. And, yes, there was a relative from Shanghai, a nonsmoker, who succumbed to lung cancer.

Such cancer stories are a poignant reminder that serious diseases in a family are a site of reflexivity about the biology of belonging.

My family's vulnerability to cancer suddenly illuminated that our kinship can be symbolized as a shared cluster of genetic disease syndromes, among other phenomena.[1] Biomedical technologies have recast the family as a biological condition, as a substantial coding of genetic risk and hoped-for customized intervention.

Especially in cancer treatment apparatuses, kinship is remade as different embroilments in the medical establishment. With the identification of cancer as an inheritable form of disease, family bloodlines are no longer just anthropological symbols but become materialized in clinics as potential genetic dangers to as well as "lifelines" for afflicted relatives.[2] As a sometime observer of family medical encounters, I am acutely aware of the ethical conundrums as an arsenal of debilitating interventions remakes the biology of vulnerability, love, and survival. As a cancer patient herself, Lochlann Jain, in *Malignant*, her view of the American cancer machine, gives a searing account of the reworking of family and gender relationships. Because I am merely a relative of cancer survivors, my approach can only be different, viewed through the lens of the new oncology, with its murky promises of genetic biomarkers and precision medicine now focused on, collected from, and imagined for populations of non-European ancestry. In the midst of complex desires of rescue and survival, could compatible genetics and shared mutations (the biology of family as a potential for life) be the techniques that can, if not unlock new "cures," at least limit cancer's malignant ferocity?

There are as yet no cures for cancer, and the disease is being managed as a chronic condition. Nevertheless, in medical and anthropological circles, narratives of cure and healing are common. Angling for necessary funds, medical researchers often invoke "finding a cure for cancer" as a major goal. In medical anthropology, there is often too stark an opposition between disease and cure, or ill health and healing, almost synonymous with another favorite opposed pair, oppression and resistance. Instead of putting prognosis and chronicity at odds, the new oncology tends to conceptualize the disease as chronic, to focus on therapeutics as both prophylactic and deterrent, and as the management of chronic symptoms. Cancer in Singapore is imagined as a smoldering fire, and the aim is to find the right kinds of drugs to control the burning.

Can customized medicine respond to a situated demand for an Asian landscape of cancer biology, discovering a new vector of risk and probability that targets a different category of global subjects? Can an alternate pharmacogenomic regime drawing on the medical constitution of family histories and ethnic-aligned vulnerabilities become a reservoir brimming with therapeutic possibilities? The new affluence in Asia has been accompanied by an epidemic of cancers, a smoldering fire in the body politic that demands urgent action and fuels pharmaceutical interest in "Asian" forms of cancer. At the broadest level, therefore, cancer research seems to lend itself to a scientific reframing of the matter of "Asia" in terms at once tragic and hopeful.

In the United States, breast cancer research has shifted from the view that it is mainly a white woman's disease to the recognition that it is a lethal threat to communities of color.[3] Nevertheless, a popular U.S. assumption is that personalized treatment for cancer depends on a highly individualized approach, traced through individual bloodlines and culminating in a fraught choice made over one's own body.

The actress Angelina Jolie's experience, published in the form of a coming-out story in a *New York Times* editorial, of preemptive breast cancer surgery highlights her family history and the inheritance of a genetic defect, BRCA1, as well as her capacity to pay for the sequencing of her genome.[4] Genetic inheritance and her personal wealth influenced her decision to intervene in a biology firmly situated in her individual body, a celebrity modeling individual initiative and resolve against a cancer residing virtually in serial bodies. Here, personalized therapies seem to translate into cancer treatment for wealthy individuals only, thus foregrounding individual self-knowledge, capacity, and decision in the American medical context.

Genetic Diversity in Cancer

> Cancer is always mutating; it is a movable target.
> —SIDDHARTHA MUKHERJEE, radio interview (2015)

Billions of dollars have been invested in the development of cancer drugs, but only a few, Herceptin and Gleevec, have been widely successful. The drugs are so exorbitantly priced that the majority of cancer patients, unlike Angelina Jolie, will never be able to afford them. Developing states facing rising rates of cancer among their populations have two options to lower costs. One is to negotiate drug prices with global corporations such as Genentech, which, incidentally, has a lab in Singapore. The other is to develop capacities in basic research in order to choose the right targets and develop more targetable therapies.

Singapore has taken a lead in emphasizing the collective approach to personalized medicine by using its genetically diverse databases drawn from different racial and ethnic patient groups. The pragmatic dimension of doing clinical trials on different groups of patients is to develop drugs that are oriented toward biomedicalized ethnic collectivities, not individuals. Oncologists and cancer pathologists note that only a few cancer genes have been discovered, and there is an urgent need to identify other kinds of cancer genes in diverse populations so that targeted therapies are oriented toward groups with high rates of the same kinds of mutations linked to particular cancers. The anticipated doubling of cancer patients worldwide by 2025 has made this field potentially a lucrative one in the Asian bioeconomy, but the larger goal is to eventually lower the overall costs of cancer treatment for huge swaths of patients in Asia.

Indeed, in the United States, the mounting costs of cancer treatment for individual patients have caused some U.S. scientists to undertake similar approaches by building collaborations between clinical research groups and enrolling patients from different ethnic groups. A *New York Times* article reports in April 2015 that the Cold Spring Harbor Laboratory in Long Island, New York, recently connected with the North Shore-Long Island Jewish Health System to gain access to "a genetically diverse patient base in the millions with different racial and ethnic backgrounds." Over a hundred million dollars have been invested in the collaboration to conduct research on human subjects that can answer questions as to why a drug that fails for one disease is actually effective in another. The stated goal is to expand cancer drug trials beyond white and affluent patients in order to "look at other racial and economic groups."[5] Singapore has moved further in this direction, taking a transethnic approach in the cancer research and therapeutic markets in Asia.

In recent years, research on the causes of this dreaded disease has focused on the genetic anatomy of cancers (actually a class of different diseases) associated with the abnormal proliferation of cells that can invade tissues and cause death in the patient. The new genetic approach is to discover the "pathways" of mutant genes and understand how they function in cellular physiology, leading to malignant transformation and progression. According to Siddhartha Mukherjee, there are three new directions in cancer medicine. The first is the hunt for therapies that target the mutant genes without killing healthy cells. In the second shift, the Petri dish becomes the context for studying how chemicals stimulate or mutate genes, thereby identifying preventable carcinogens (environmental and lifestyle causes) in the lab, instead of depending on epidemiological studies. A third technique is screening for different kinds of cancer genes.[6]

In the Singapore research milieu, the genomic approach to cancer set off a hunt for a wider spectrum of cancerous genes than those already found (e.g., the BRCA1 and BRCA2 genes for breast cancer) in Western cancer studies. In the hope of developing an exquisite specificity for targeting cancers, researchers in Asia look at Asian populations assumed to have different genetic variants and mutations.

Ethnicity in targeted cancer therapies, therefore, is an effect of the genomic approach, which searches for different kinds of cancer genes, as well as the brand of an emerging research milieu that ought not to be subsumed in a Western critique of racial stigmatization or biologics. By deploying ethnicity as a code for identifying specific genetic risks and mutations, cancer research in Singapore may be considered an implicit and sometimes explicit criticism of the presumption in Western medicine of a genetically uniform species as an alibi for an assumed whiteness. This form of a genetically driven approach to cancer therapeutics requires extreme specification to the internal Galapagos of cell attack and resistance. The goal is to customize the right cocktail of drugs for a particular patient so that the disease can be managed as a chronic condition. Nevertheless, by zeroing in on ethnic, family, and individual genetic targets, the new research milieu is also productive of affects of fear, hope, and pride in a novel form of scientific self-knowledge, one that frames "personalized medicine" as situated within a diagnosis and affective formation of collective bodies. I would argue that in the Asian biomedical frontier, a new ethnicization marks affective aggregates emerging from a complex environment. The biomedical optics I present here are kaleidoscopic, with "ethnic" or "ethnicity" as one of many components in shifting combinations. In Singapore's Biopolis, we encounter a different patterning of elements as DNA, mutation, family, identity, geography, and ethics are assembled and enacted in a situated strategy of therapeutic intervention.

Contrary Affects

In medical anthropology, the recent focus on "technologies of care" identifies affect as the subjective experience of well-being among patients subjected to different kinds of clinical intervention.[7] In contrast, I do not dwell on physician–patient encounters or on the subjective experience of patients.

Rather, following Deleuze on Spinoza, there are two other ways to think about affect: first, as the way interacting things affect one another, and second, as the experience emerging from such encounters. Affect refers to the human capacity to affect and be affected and is thus the capacity to shift from one experiential state to another. Affect is not subjective feelings, but an abstract intensity in the passage from one experiential state of a body to another.[8]

Because affect is transmittable in ways that feelings and emotions are not, it is potentially such a powerful social force that it precipitates further action in a dynamic field of relationships.[9] Therefore, instead of viewing affect as an externality to science (as in many anthropological accounts), affect is integral to the realm of verifiable scientific claims and to a new set of affective attachments to science objects, techniques, and programs deemed, rightly or wrongly, to be beneficial to health.

Specifically, I argue that the cancer experimentation in Singapore generates the experiential condition of "contrary affects" that strengthen affective investment in the cruel science itself. By "contrary affects," I mean a mutually productive relationship of vulnerability and hope that is attached to a problematic condition, such as susceptibility to fatal diseases. In a somewhat similar vein, Lauren Berlant notes that "a relation of 'cruel optimism' exists when something you desire is actually an obstacle to your flourishing."[10] Thus, vulnerability is also a condition of hope, and vice versa, for populations threatened by a spectrum of cancers. The new oncology, however, is not just about transforming a preexisting debilitating condition into hope. It is productive of both that sense of hope and sense of vulnerability in the same moment, which reflects in its very framing a sense of collective cum personal vulnerability that it aims to resolve by offering "hope." My diagnosis discloses the paradoxical attachment to genetic risk for cancer, as research on such a compromised condition engenders hope for better medical outcomes. The making of hope depends on the imaging of vulnerability, associated with the very condition of being perceived differently, as the ethnicized cancer patient, which is to say, to be alive "in this way."

Affective attachment to an injured state is connected to the promise of improvements in the future. Research on cancer transmits affects that are a perverse mix of genetic risk and therapeutic hope, because the violence of disease now can be suspended by the promise of novel technologies. Apart from recoding a genomicized cancer as a site of medical hope, then, cancer research links debility and potential rehabilitation, such that each cannot but evoke the other. Because genetic risks come to bear ethical values, they prompt scientific efforts to design new solutions for this set of terrible diseases and, in the process, expand relations into the future.

The Biopolitics of Deferral

Here, the tensions of the biopolitical are revealed, in that it is no longer merely a governmental intervention to secure a maximized, flourishing population, but, instead, enacts a kind of contingent optimism that is part of being modern. First, the biopolitical strategy subdivides the population into groups that do not follow the macro-sociological fault lines of ethnic, nation, and so on, even as they build on them. Rather, genomic research recodes such loaded macro-demographic specifications in relation to making more of life, to enact a new political wager on collective life menaced by fatal diseases. As earlier chapters show, remnants of biopolitical forms and modalities of governance and regulation are refurbished, reconfigured, and rearticulated through biomedical research apparatuses and public health, including ethnicized interventions into contemporary cancer genomics as disease research.

Second, a common misreading of biopolitics as just "power over life" or a disciplinary control over hope has turned anthropological attention on the politics of sheer survival that is perhaps more inspired by Giorgio Agamben's formulation of "bare life" or *zoe*-politics[11] than by Michel Foucault's concept of biopolitics that is about the modern government of well-being. In *Security, Territory, Population,* Foucault defines "biopolitics" as an originary European style of government that operates at the confluence of life and politics centered on the security of the population. An array of life-improving techniques intervenes into the geographical-climatic-physical milieu within which the human species can be protected and ideally flourish.[12]

The analytics of "hope" that I identify are centered not on bare survival,[13] but rather on the Foucauldian biopolitics of improving life and a style of living for a reason. It has to do not simply with living, but with living better in the future. Indeed, Hiro Miyasaki has proposed a method of hope that depends on projecting a specific temporality of the future and deferral that allows one to survive an intolerable present.[14] Biopolitics in Asian sites have turned to biomedical science as an expanded source of hope for populations troubled by the question of aging and dying with dignity, of preventing the previously unpreventable. The development and application of biomedical techniques engender anxiety and hope within a zone of deferred death enabled by novel medicine, which is perhaps another site of the contrariness.

I view Asian oncology as a pharma-neoliberal intervention that configures a biopolitical space of deferral. The genomic suturing of dread and optimism is a kind of attempt to recuperate vulnerabilities, which already registers the possibility of death, rather than erases it.[15] Such a biomedical technology of deferral animates contrary affects that simultaneously maintain and underscore the precarity of the present-future. Contrary feelings of fear and hope fuel a sense of existential apartness, or exceptionalism.

In science and technology studies, "genetic exceptionalism" has been identified as generative of new biosocialities or social networks based on a sense of affliction that can press for special biomedical attention or fight discrimination or exclusion by insurance markets.[16] Genetic exceptionalism thus has been conceptualized as a mode of shared affliction that converts into stimulus to social action, and then reclaimed social experience, among those identified as suffering from genetically compromised diseases. The trajectory is from exceptionalism to biosociality, or, affectively, from victimhood to affirmation.

By contrast, the kind of biosociality I identify as emerging is not based on genetic exceptionalism of affliction, but rather on "ethnicity" as a biomedical category of exceptionality in the sense that different ethnicities have become specially disposed/vulnerable. In this context, where cancer research generates both hope and vulnerability, what emerges is actually a sense of the ethnic group as susceptible to cancers in rather specific ways. Here customized medicine generates the health phenomenon in which, as Joseph Dumit has observed, pharma has increasingly switched from treating illnesses to treating illness risk, thereby putting all potential sufferers of diseases (ethnically differentiated in sampling) into a captive customer pool for chronic treatment.[17] It is not exceptionalism but the generality of cancer diagnoses that links this new "biosociality" and that is generative of the affects of dread and hopefulness.

Furthermore, if accounts of biosociality view genetic exceptionalism as part of a new mobilization through which communities are developed, often against powerful state and corporate actors, the identification of genetic differences in Asia entertains a more complex relation among scientists, populations, and state power. It is the community of research scientists, not grass-roots groups, that affects the biosocial condition of genetic solidarity. Here the intervention is not merely at the clinical level, but at the level of the state and medical elites actualizing techniques of care aimed at groups deemed specially vulnerable in society. In Singapore, physicians and researchers also come to recognize a state of exceptionality in the experience of themselves as an ethnic group particularly predisposed/vulnerable to virulent forms of cancer. The kind of genetic exceptionalism I identify is an affective attachment of researchers to diseases, bodies, and life chances of peoples in the region, albeit in selective focus on a majority population based on pragmatic and cultural claims. More than any other medical field, the new cancer experimentation engenders contradictory affects emerging from the identification of selectively "Asian" biological risks and strategic opportunities in this biopolitics of disease management and deferred death that configures a kind of ethnicized space of intervention that Brian Buchanan calls "geo-biosociality."[18]

Below, I track the assembling of two ethnically aligned entities in "Asian" cancers: the mutation and the biomarker. Because such cancer artifacts include ethnicity and gender attributes, heuristics, and markers themselves, they can affect the experiential state of the experiment and invest it with a dimension of ethnic exceptionalism. I first describe the assembling of a genetic cancer database that is, I argue, materially, logistically, and emotionally potent in triggering pharmaceutical and therapeutic hopes. I then turn to the creation of Asian cancer biomarkers that combine biological properties with racial and gender attributes. Both the cancer DNA map and the cancer biomarker thus straddle scientific and symbolic differences, at once exerting powerful affects not of victimization but of scientific optimism and possession of the "Asian cancer problem." Cancer research under way in Singapore, I suggest, engenders a state of passionate attachment to a disease that comes to be imbued with both dread and hope.

A Hunt for Asian Biomarkers

The new oncology is dominated by the study of signaling pathways that tracks the flow of genetic and protein information involved in the emergence and expression of the disease. In this molecular approach, the focus is on thousands of genes in an individual patient (thus, it is a highly personalized form of medicine). Nevertheless, characteristic aberrant genetic or molecular elements are often coded as biomarkers that can come to acquire other (ethnic) diacritics as well. By "discovering" biomarkers for serious diseases, researchers and clinicians add value to their knowledge of disease medicine, and they increase their capacity to match patients to clinical tests for appropriate treatments.

Because biomarkers can accommodate diverse viewpoints and interpretations, they have been theorized as "boundary objects."[19] The novel Asian cancer biomarker is an especially rich and ambiguous boundary object in that it inhabits intersecting material, social, and semiotic worlds. As a measure of the risk, drug response, and drug target of a disease, biomarkers may come to bear ethnic or gender features as well. As a particular condensation of genetic, epigenetic, and cultural elements associated with Asian forms of cancer, the biomarker fuels affects of how ethnicity, gender, and "Asian" are at stake.

An oncologist named Dr. Lin noted that after DNA and lifestyle are taken into consideration, "family history is still important." He explained that germline cells disclose both genetic and epigenetic effects on the expression of a disease in a particular patient. Because patients inherit DNA and cultural elements from their families, their "ethnicity" (consisting of entangled biological and cultural lifeworlds) is also picked up in the creation of protein biomarkers. So when researchers invoke the so-called South China founder effect in distinctive forms of cancer, they integrate genetic and epigenetic inheritances. The assumption is that since such populations historically evolved in the same ecosystem, their somatic mutations or spontaneous errors in DNA replication can induce high rates of vulnerability to certain diseases. That is, the etiology of these specific cancers is linked with a complex of inextricably linked factors that speak to forms of human collectivity over time and in specific ecosystems.

Therefore, for Asian researchers, "ethnic" or "ethnicity" is this cluster of interacting genetic, epigenetic, and geographic elements that shaped a people and their inherited medical chances. In this scheme, my family is descendants of peoples who evolved in what is now South China and are thus inheritors of the entangled DNA and cultural legacies dubbed "ethnic Chinese." This cancer assemblage of somatic mutations, biomarkers, and family cell lines performs and illuminates what is at stake for being "ethnic" beyond a "biologistical" construction of racial category, as Duana Fullwiley has argued.[20] In the new oncology, "ethnic" or "Asian" is a reassembly of experimental heuristics and therapeutics in a distinctive geo-biomedical-biosocial complex.

Oncologists in Asia focus on a cancer-linked somatic mutation (the epidermal growth factor receptor, or EGFR) that is often associated with aggressive forms of breast and lung cancers found in Asian populations. The biomarking of this EGFR mutation as "Asian" and its successful treatment with various EGFR inhibitor drugs has identified a broad area of health vulnerability and, by extending life, given hope to patients suffering from such cancers. Researchers point to their discovery of biomarkers such as the "Asian female breast cancer" patient, perhaps also aware of the emotional potency of such hybrid science objects. While these biomedical objects may sound disturbing, they generate an affective resonance akin to racial pride, in both the perverse sense of possessing at-risk bodies and identities, and the benign sense of attracting biomedical expertise to Asian forms of deadly, inheritable diseases.

"The Female Asian Nonsmoker"

The finding of somatic mutations in Asian cancers has enhanced the cultural aura of biomarkers. For instance, Asian oncologists were also among the first to match mutations in lung cancer to precision medicine. Their pivotal testing of a lung cancer drug among patients drawn from the region created new bio-markers with specific ethnic and gender associations that register the cultural stakes of a scary disease that plagues many in Asia. By naming a biomarker "female Asian nonsmoker" and the drug "Iressa," physicians offer hope for converting a deadly malady into a chronic one, while creating distinctive can-cer biomarkers that can promote research and novel therapies.

By tracking biomarkable mutations and testing drugs, we will see that these science practices are also steps in layering social features onto what are presumably molecular biomarkers. A decade ago, a lung cancer drug named Gefitinib tested poorly among Caucasian patients, and the U.S. parent com-pany was on the verge of folding when it turned to testing in Asia. The drug targets a somatic mutation in the so-called EGFR receptor, a condition found to be more pervasive for Asians than for other populations. An initial study indicated that the lung cancer drug seemed to improve the survival rates of nonsmokers in Japan and other Asian countries. Post hoc studies found that there is no difference in the efficacy of the drug for "Chinese ethnic groups" and other Asian populations. In 2009, a Hong Kong oncologist, Dr. Tony Mok, ran a successful "pan-Asian" clinical trial for the anti–lung cancer drug, now called Iressa.

It is instructive to see how the clinical trial links the EGFR mutation, Asian women, the region, and the drug within a chain. In a 2009 interview, Mok says, "You know that EGFR mutation is actually more common in East Asia, espe-cially in the Chinese population," thus quite explicitly collapsing geography and population categories. He explains, "We have to know whether they're mutation-positive or mutation-negative. So I think this is an example of what we call personalized therapy."[21] His report identifies a geographical/racial cat-egory, "pan-Asian," from which 1,200 patients were recruited. Patient groups recruited in Hong Kong, the People's Republic of China (PRC), Thailand, Tai-wan, Japan, and Indonesia collectively are referred to as "East Asian." About 60 percent of these East Asians are found to have the molecular biomarker that is said to be inhibited by Iressa.[22] The Iressa test connecting the EGFR molecular biomarker and East Asian subjects was one of the first successful treatments for cancer therapies in Asia. It fueled demand for more testing techniques that result in the bundling of a mix of indicators—molecular and sociocultural—within cancer biomarkers.

In Singapore, the "East Asian" biomarker for lung cancer acquired gender connotations as indicated in the drug's name change as it went to market. Dr. Kwong, a physician from a leading hospital, observed that the drug Iressa was especially "effective on Asian female nonsmokers. The ethnicity of the patient matters: that is, the EGFR (protein, or HER 1) mutation is higher in Asian women."[23] In other words, such findings about somatic mutation and effective drug testing have allowed oncologists to make "the female Asian nonsmoker" a biomarker for the drug Iressa. This particular biomarker is thus a configuration of molecular, ethnic, and gender elements that has been invented by pharmacogenomic testing and is reproduced by the prescription of successful drugs.

This pivotal trial illustrates the different ways in which ethnicity or gender variously operates as a sampling tool, a genetic fingerprint, and an experimental subject. While gender has remained a stable signifier, the elastic ethnic indicator oscillates among Chinese, East Asian, and pan-Asian, thus expanding the symbolic power of racial risk profiling beyond the laboratory. By naming the "ethnic" and "gender" indicators of lung cancer, physicians and researchers perhaps hope as well to promote the recruitment of patients as experimental subjects in future clinical trials.

By comparison, there is a dearth of racial diversity in U.S. clinical trials for cancer, despite the links between ethnicity and biological response to drugs. Oncologists attribute low participation rates to both mistrust of health trials by African Americans and ignorance of the value of the tests for patients. Only recently have researchers at the University of California started a program to recruit minority groups because "the two fastest-growing minority groups—Latinos and Asians—are experiencing cancer as the leading cause of death."[24] By comparison, Singapore has had a multiyear head start when it comes to the careful matching of Asian patients with clinical trials and cancer therapies.

Cancer, Kwong remarked, "is like a smoldering fire that occasionally needs to be extinguished." Because lung cancer was a common cancer in Singapore, Iressa was presented to the public as an object of optimism for a tragic or toxic condition, accidentally tailor-made with ethnic specifications. In addition there was the urgency of finding an effective drug, given the finding that lung cancer was the second most common form of cancer in Singaporean men and the third most common in women. Cancer rates are rising throughout Asia as well. Kwong noted that Iressa promises the cancer patient hope by "converting her lung cancer from a killer disease into a chronic condition that can be managed."

The physician then raised a question of whether the government should consider making personalized medicine more available for those who can benefit from it. Is the prolonging of patients' lives (by two years in advanced cases) worth such expensive drugs (more than US$1,500 for a month's supply)? He suggested that perhaps the size of a tumor (one cubic meter carries one billion cells) could be a threshold for access to subsidized medicine. Iressa therefore is a cancer drug that makes the relationship between the public coffers and the citizen's body part of a biopolitical calculation that balances two kinds of public good: health and solvency.

However, the framing of cancer research beyond its status as an individualized disease makes possible other ways of conceiving the necropolitical problem of funding, as individual incidences become legible as the specific actualization of a generalized condition of vulnerability. Indeed, by making visible the compromised DNA of a particular ethnicity, and by making vivid the ravages of the disease, research physicians are hoping to pool patient risks by promoting gene testing and insurance coverage. Ethnic or genetically proximate groups that share the same mutations can be tested for a few innovative drugs. This group approach can be covered by insurance, thus lowering the cost of expensive medicine. Indeed, a few months after Kwong's talk, Iressa was approved and covered by insurance companies for patients in Singapore. An online health website invoking "biological drugs," "female, Asian," and high cancer death rates helped convey a new compassionate policy that provides "MediSave" subsidies for buying Iressa at half its U.S. market price.[25]

The Iressa test highlights the female signification of customized therapies. A cancer center website notes that "drugs such as Iressa, Tarceva, and Erbitux are biological drugs which have been proven to work in patients with the EGFR mutation. These patients are most likely to be female, Asian and never smokers." The female cancer biomarker in turn reconfigures the therapeutic and insurance environment responding to a collective ethnic insecurity. The creation of biological-gender biomarkers forces a rethinking of kinship and ethnicity, thereby reinforcing a group approach to medical insurance and affordable drugs. It is a pragmatic translation of medical social policy that can be contrasted to the U.S. health situation, where extremely costly and individually oriented genetic diagnoses are available only to patients such as Angelina Jolie and Steve Jobs. In comparison, the "Asian female nonsmoker" biomarker reassembles personhood into an individual appropriate to epidemiology as part of a scheme to target and economize medical insurance.

The implosion of informatics, biologics, economics, and culture gave rise to the creation of the nonsmoking Asian woman as a condensed icon that inhabits an emerging terrain of biomedical science. I would argue, however, that the "Asian female nonsmoker," unlike the Oncomouse, is not merely a patented hybrid object "enterprised up."[26] As a creature of genetic technologies, the Asian female nonsmoker circulates in a material-semiotic field of affective bond between the cancer expert and the oncology of ethnicity itself.

The Heart of the Matter

In her book *Malignant,* Lochlann Jain describes the U.S. cancer milieu as rife with conflicting goals of denial, profit making, and cure seeking, creating a quicksand of uncertainty for patients and doctors alike. Singapore's cancer milieu is also market-driven, but perhaps tempered by a situated goal of attending to the special needs of Asian patients, both to help them and to build a new pharmaceutical space.

In this public-private world of modern medicine, the clinician-scientist, inventor of hybrid biomarkers, is himself a kind of boundary subject. He is a risk-taking researcher who is poised between seemingly objective scientific work and the ethical ambitions to bridge cancer fears with medical promises. For oncologists I spoke with, desires to advance one's career in this fast-moving field are more likely than not fueled by hopes to improve the survival rates of patients with "Asian forms of cancer."

I met Dr. Ang, a leading cancer specialist, within the high-tech environment of the Duke University medical school. He first offered to make me a cappuccino from the fancy coffee machine in his department's pantry. In appearance and spirit, Ang reminded me of my brother, the professor of physics at Princeton who donated bone marrow to our older brother, who was a surgeon in Singapore. I recognize in them a fraternity of unassuming, brilliant, and dedicated professionals one encounters in overseas Chinese families that have been maligned for being capitalist nerds but who constitute the backbone of modernity in Southeast Asia.

Ang's pale, owlish face masks a passionate engagement with the larger implications of new cancer biology in Singapore. As a student, Ang was sent by the Singapore government to train "in a typical oncology lab in Stanford." In Palo Alto, he saw that breast cancer patients were women in their fifties and sixties. When he returned home, he learned from anecdotal stories and his own observations that breast cancer patients in Singapore tended to be women in their thirties. Indeed, ethnic Chinese patients have a propensity to experience an early onset kind of breast cancer, rather than the middle-aged form more prevalent among patients in Western countries. Ang had an epiphany. "So the question posed itself: Why not use techniques, knowledge, and philosophy surrounding breast cancer treatment and apply them to Singapore and the region?"

In a quiet way, professionals in disease science reveal a deep appreciation of what it means to be staked collectively as "ethnic" or "Asian." As researchers in a cutting-edge field, their scientific conduct is deeply intertwined with biological and cultural variations that help capture the specific manifestations and inscribe the affective significations of their findings. Ang suggests that we think of biomarkers not as mere bioindicators but as hybrid objects that can and should index cultural and social differences. He said, "A biomarker is a collection of measurable biological features, that is, an intentionally loose definition. One example is a nonsmoking young Asian woman with lung cancer who responds well to the drug Iressa."

But when I asked, "Is this an effect of science reinforcing notions of Asian ethnicity?" he shifted to his neutral physician pose and gave an appropriate response. "This is not a big deal for me. I have a patient who has cancer. I leave it to others to deal with ethnic links to predisposition for the disease. My technical know-how is focused on managing this individual's problems." But this know-how designs the ethnic biomarker, one that crystallizes an affect of genetic exceptionalism when it comes to vulnerability to fatal diseases and to intellectual capacity to take on the problem. In the process of hunting for Asian biomarkers for cancer, researchers are also exercising representative powers that reanimate affects associated with older forms of collective vulnerability and hope.

"Our Cancers, Ourselves"

The search for Asian biodata, samples, and biomarkers in Singapore aims to transform the uncertain outcomes of cancers into predictive and economic values on the one hand, and moral affects and facts on the other. New expertise in risk diagnostics, prognoses, and drug discoveries shapes a research infrastructure that is entangled with but not overdetermined by a process of biocapitalization. Rather, the search for Asian cancer mutations and biomarkers engenders a broad therapeutic hope that, while logistically linked with big pharma's market concerns, is a material and affective "investment" in a novel "Asian" collectivity.

Oncology as productive uncertainty in Singapore may be said to do three complexly interwoven goals. First, it is a strategic attempt to be a competitive biomedical hub against other ones (e.g., in Bangkok) by building an extensive yet Asia-oriented knowledge and infrastructure for disease science, with a focus on a global, runaway deadly disease. A leading local epidemiologist Dr. Lau noted, "Biomedical finds are like software; money is to be made through the whole value chain." This research milieu with its existing and anticipatory forms of governance and aggressive promotion of ethnically aligned DNA and diseases has been catalyzed by shifts in the pharmaceutical industry and projected Asian population markets. Because of aging populations, aging drugs, and rising costs of drug development in the West, the moment is ripe for the growth of health markets in Asia.[27] Even though new drugs for diseases prevalent in the region are still many years away from development, Singapore's demographic diversity is offered as a scalable genetic microcosm of the vast continent's populations, especially in the explosive growth of cancer as a major focus of big pharma.

Second is the political opportunity to gain some kind of sovereign or regional control over biomedical data, objects, and samples, and that in a powerful way as well accounts for the naming of things in racial and ethnic terms. The ethnic coding of DNA and disease expression, I argue, cannot be reduced to biologistical constructions of ethnic and genetic exceptionalism that registers a double sense of distinctive bodies and geosociality. By recoding ethnicity as a horizon of potential debility and rehabilitation, scientific practices create new DNA maps and diagnostic objects that are infused with the affective admixture of vulnerability, protection, and optimism. The community of Asian scientists deploys ethnic/ethnicity as a specific coding and territorializing device that, through difference and repetition, configures a distinctive space of cancer research and intervention, thereby shaping a sense of sociality from above. New cancer research remakes the terms of the general and the exceptional in genetic, cultural, ethnic, and geographic belonging: a model of collective "biological citizenship" different from those emerging from grass-roots organizations.[28]

Third, an affective assemblage of ethnicized risk, probability, and potentiality for remediation builds social expectations of novel medicine. I have framed the affects thus engendered as a paradoxical simultaneity of dread and hope, vulnerability and optimism, that is, contrary affects that infuse the scientific objects and endeavor with moral and psychological mastery of "Asian bodies" and the problems therein. "Genetic exceptionalism" has less to do with biomedical segregation or victimization and more to do with a situated reconfiguration of knowledge-making practices and a collective experiment oriented toward their world. It aligns the sociocultural positions of experts and would-be patients, while animating professional pride in a science worth doing, and a science that grounds Asian life as worth living.

Fourth, the implosion of science and culture, subjects and objects, in the new biotechnologies has created chimerical objects that are both lab devices and proprietary objects of biocapitalism.[29] I would add that the "Asian female nonsmoker" is not just a diagnostic artifact but a new cultural figure who embodies the experience of being both Asian and female. By identifying "the norms and forms"[30] of finitude, Asian oncologists also study unconscious representations of how Asian patients represent their conditions to themselves. The medical conditions of possibility for survival also animate specific cultural representations by which a culture configures meanings of gender, life, and death. It is this interplay of tangible and intangible elements on the one hand, and of productive uncertainty and hope claims on the other, that crystallizes a new field of "Asian" cancers and the complex of geo-biomedical-sociality.

As an emerging knowledge system, the Asian oncology that Biopolis plugs into vividly illuminates what is at stake for local scientists reconfiguring Asian bodies as subject-objects in a resolute quest for psychological, moral, and geographical knowledge of genetic identities and frailties. Thus, the sense of exceptionalism is about not an individualized self, but a "self" as an ethnicity being formed through self-knowledge and through its co-ethnic scientist avatars. As Ang said, "It is a question of building up a new database, of funding them and *moral and psychological ownership* in setting it up" (my emphasis). Here is a scientist's articulation of how a constellation of cancer bioindicators can itself be a process that is saturated with emotional belonging, providing a scientific map that "tells us who we are" through a repositioning of the core of our beings. If it is not essentialist, then the core is relative to being "Asian," in that it can be realigned. Through the tracking of Asian cancers and the discovery of Asian biomarkers, the space/geometry site and location of the core or soul is, in the mind of the oncologist, projected as a geo-biosociality.

My relatives are long-term cancer survivors, their treatments evolving through decades of cancer medicine and the rise of genomics research. The mystery of one's fate is cast by the random dance of biological and cultural factors, a throw of the die that, by sheer dumb luck, does not fall in one's corporeal field. My older siblings, unfortunately, might have been exposed as infants to damaging chemicals during World War II in then-Malaya. But new cancer therapies make explicit DNA as a medium through which family ties are actualized and reconfigured. Our mixed feelings about surviving are like the smoldering fire that the science of DNA can only douse now and again, but to our abiding hope and gratitude.

PART II

UNCERTAINTIES

FOUR

THE PRODUCTIVE UNCERTAINTY OF BIOETHICS

Dr. Mega, a Cambridge-trained Malaysian geneticist, expressed skepticism that bioethical norms were by themselves effective in protecting the interests of native groups. She said, "We have enough of 'parachute research' by outsiders, and thus we are reluctant to send local samples overseas." Her testimony underscores a more pervasive doubt that casts the securing of informed consent as a technical mechanism, albeit with an ethical gloss, for the capture of genetic materials. The uses of "informed consent" may permit foreign researchers to bypass a more genuine and caring mode of exchange that is meaningful and sustainable for vulnerable indigenous communities who have come under the molecular gaze of researchers worldwide.

Meanwhile, in Singapore's gleaming labs, bioethical norms are also viewed as inadequate in a different way. Bioethical "best practices," scientists argue, are merely a first step in the island's rise as a center for clinical trials, not a sufficient condition for the achievement of a world-class status. A U.S.-educated oncologist, Dr. Tan, quipped, "Doing clinical trials is like trying to be a franchise to a global system like Starbucks." Some may cringe at the idea that ethics and franchising go together, but best practices of "ethical capitalism" are now normative in international business, including global drug companies. However, implementing the bioethical procedures of informed consent alone does not attract pharmaceutical investments.

It is only one threshold among others, Tan noted, that make labs even eligible for consideration. He invoked "scientific entrepreneurialism," not the kind that gets around bioethical norms to make bodies and tissues available for pharmaceutical plunder, but rather how the demonstration of bioethical rectitude has become the very condition of market competitiveness. In other words, bioethics is part of the uncertainty confronting reputable science research, but it can be a productive element in shaping a biomedical platform that mobilizes experts, information, bodies, and capital for running clinical trials.

This chapter illuminates how researchers at Biopolis not only challenge the figuration of the non-Western scientist as a complicit or exploitative agent of biocapitalism but also show up the "productive uncertainty" of universalistic bioethics as merely one element in the making of reputable bioscience. The outline is as follows. I first trouble anthropological framings of biomedical operations and actors in non-Western countries as uniformly lacking in the enactment of bioethics. I also show that researchers in Malaysia and Singapore are moral problem solvers who view international bioethics as at best uncertain and inadequate guidelines in response to the needs of experimental subjects who contribute blood samples.

The larger implication of the productive uncertainty of bioethics is that the implementation of bioethics alone, without surrounding infrastructural and cultural capacities, cannot build reputable science anywhere. I will illuminate how science leaders in the Biopolis complex install regulatory and infrastructural regimes, as well as draw on cultural skills, in making Singapore a global biomedical platform for clinical trials.

Center-Periphery Bioethics?

Anthropology has long been skeptical of modern medicine in non-Western environments, viewing medicine as a tool consistently deployed for the plunder, exploitation, and neglect of native populations. This framing of medical imperialism extends also to a suspicion over the place of anthropological research in furthering the exploitation of erstwhile subjects, as in an infamous case of anthropological complicity in a controversial experiment to test Yanomami susceptibility to measles in the Amazon.[1] In postcolonial, socialist Venezuela, Charles Briggs and Clara Mantini-Briggs document the ways in which medical authorities not only neglect to help aboriginal peoples in a cholera epidemic, but indeed blame them for spreading it.[2] Unequal power relations between medical experts and the vulnerable populations among whom they conduct research have become further entrenched, it seems, by the growth of the global pharmaceutical industry, which prowls the world for indigenous samples and knowledge as resources for commercialized medicine.[3]

More recent accounts address the deceptive, manipulative, and at best partial application of bioethical norms, whether in the collection of samples and other bioresources from indigenous communities or the treatment of experimental subjects in poor countries. Kaushik Sunder Rajan offers a broad framing of the "experimental potential of Indian populations as trial subjects" in clinical trials that evade regulatory constraints in advanced countries.[4] Adriana Petryna describes another concept, "ethical variability," in the global spread of clinical trials that circumvent U.S. regulations by conducting tests on "treatment-naive" native proxies in the global periphery.[5] The picture of biomedical experiments in emerging regions has been dominated by such issues as the availability of surplus populations for exploitation and their vulnerability to unethical treatment in clinical trials outsourced by drug companies from the United States and Europe.

While such critiques are important for locating serious violations and lapses in the global operations of biocapitalism, they risk reinforcing outdated binary schemes reminiscent of the "world-system model"[6] that is ostensibly perpetuated by biomedical enterprises. An overly homogeneous picture suggests a single, core science-capitalist zone that exploits global peripheral areas, divided into "middle- and low-income countries" where unregulated science prevails.[7] The scenario of biocapitalism further entrenching global asymmetries unfortunately gives some distorting impressions of how drug companies strategize in the developing world. The emphasis on "potential trial subjects" gives the impression that clinical trials are driven mainly by the quest for plentiful exploitable experimental subjects, while ethical variability implies that contract research organizations (CROs) working for drug companies can be careless about upholding ethical norms and quality controls. But while unethical operations may provide a seemingly alluring cost-cutting loophole, reputable drug companies are deterred from investing in poorly regulated trials if the resulting drugs fail to win global approval. The U.S. Food and Drug Administration (FDA) acts as a global authority to award licenses to "non-inferior drugs," which are required to be produced under highly regulated conditions.[8]

There is some evidence that trials that neglect the welfare of participants have suffered from chronic inability to win global approval for the drugs.

Indeed, the Indian biomedical enterprise Ranbaxy, which was the subject of Sunder Rajan's ethnography, has been repeatedly banned from selling drugs in U.S. markets.[9] For globally successful drug companies, trials in "treatment-naive" situations are a bad investment indeed because drugs tested under questionable standards fail to win approval for entry into lucrative markets. In other words, finding exploitable test subjects is a less likely proposition for what is still a U.S.-centric global drug industry than has been claimed. Opportunities to conduct unethical research actually drive away drug companies that need to avert claims that their products were tested on vulnerable populations, under regulated conditions.

Furthermore, claims about ethical variability in the emerging world do not capture the uneven qualities of biomedical centers across a vast region. World-class biomedical hubs in East and Southeast Asia are currently fostering a boom in hospitals catering to middle-class patients numbering in the hundreds of millions.

As it may happen anywhere else in the world, lapses in ethical practice may occur, and errant researchers and experiments are subjected to the oversight of local and international review boards. Indeed, it is in such areas of rigorous regulatory and ethical oversight that Singaporean research and health institutions are in fierce competition with these other pharmaceutical centers.[10] The North-South framework is seriously out of date as well in its broad-brush representation of non-Western regions as lacking in professional ethics and reputable scientific experiments, or as places that exist solely through the template of neocolonial bioplunder. The implication is that health professionals in emerging countries are at best unprofessional and at worst compliant tools of corporate behemoths from the West. Instead of the implication that health workers and practices in non-Euro-American sites are automatically inferior, we need to investigate how in some places good science is practiced and may indeed contribute to debates on bioethics in the making of cosmopolitan biosciences.

Problematizing Bioethics

Newly affluent Asian nations have embarked upon systematic scientific projects to improve the standards of health and well-being for their citizen-subjects.[11] State-sponsored biomedical institutions and enterprises have sprung up in China, Japan, South Korea, and Singapore. Contrary to conventional views, drug companies and CROs hope not so much to escape regulatory constraints in the West in order to undertake unethical research among easily exploited natives elsewhere, but rather to find emerging sites where bioethical norms are part and parcel of biomedical platforms. Furthermore, given the pursuit of biomedical excellence in a fierce, globally competitive marketplace, there is the possibility that scientists may be eager to treat their own people not as exploitable research subjects, but as potential beneficiaries of modern medicine in the emerging world.

According to Marianne Talbot, a professor of bioethics at Oxford University, bioethical norms devised in the North Atlantic science world are premised on "the importance, to ethics, of the notion of free will," and the capacity to morally differentiate between right and wrong. Only mature human subjects "are deemed capable of choosing freely to act for *moral* reasons."[12] Bioethical regimes regulate the moral obligation of physicians and scientists to protect the confidentiality and autonomy of patients and research subjects as free individuals.

In implementation, bioethics is productive of uncertainty because of the gap between received bioethical guidelines and the moral dilemmas researchers encounter in dealing with human subjects. Anthropologists have argued that bioethics should not be understood as an established set of universalizable guidelines, but as an evolving process of constant negotiations as bioscientists and health professionals encounter moral dilemmas in different contexts of intervention into life. Indeed, the anthropology of clinical practice has been especially alert to workplace negotiations between external bioethical guidelines and "practices of medical responsibility," to the extent that some health workers have become "moral pioneers."[13] Indeed, professional work ethics involves a constant practical engagement with lived moral problem solving. Klaus L. Hoeyer and Anja M. B. Jensen identify a mode of "transgressive ethics" whereby surgeons have to make a series of moral choices concerning the materiality and temporality of organ harvesting.[14] While possibly not as dramatic, bioscientists in the field also confront a series of often conflicting concerns and positions in an expanding and fast-changing scientific landscape. Therefore, the better question is not why (some) scientists are unethical, but what forms of ethical interventions ought to be taken in milieus of uneven or unsatisfactory regulations?

What is interesting about emerging biomedical frontiers in Southeast Asia is skepticism about first-world ethical regulations governing experiments, that is, uncertainty that the bioethical "best practices" are adequate for protecting patient interests and for building reputable research projects. Situated skepticism is particularly generative of debates on bioethics, and not just the failure of these globally circulating norms to reach the ground, as it were. Such doubts are stimulated not by the refusal of prevalent bioethical norms (embodied by informed consent forms) but by whether they are in and of themselves sufficient for protecting the moral and social interests of research subjects or for building globally competitive biomedical science. Asian skepticism is an engagement with the ready-made suite of bioethical guidelines emerging from moral and practical negotiations of the biology and the politics of biopolitics in diverse sociopolitical contexts. The uncertainty of bioethics in emerging sites of research is productive for raising questions that can both sensitize and enrich scientific responses to evolving bioethical dilemmas. I make the following observations.

Instead of looking at researchers' moral agency in terms of conformity or nonconformity to conventional bioethical instruments, I explore how scientists struggle for contingent resolutions to ethical dilemmas encountered in particular research milieus. Whether collecting DNA or running clinical trials, researchers deal with a particular mix of material, practical, and cultural conditions that calls for a situated approach rather than that prescribed by universalistic guidelines.

Clearly, by premising bioethics on the individuals exercising free will, global bioethical standards cannot take into account political conditions that shape collective action or otherwise constrain individual intentional action. Often, even when "voluntary consent" is obtained for an experiment, test subjects cannot be assumed to be actually registering uncoerced decision or moral intentionality (based on Western notions of free will). In diverse cultural contexts, scientists may need to be flexible by devising alternate ways to "compensate" experimental subjects for their participation.

Bioethical standards alone cannot reduce the uncertainty that haunts the success of any scientific endeavor. Scientific standards and technical regulations must be in place in order for a specific science outcome to be certifiable and patentable. To be reputable, scientists must be held accountable for their professional actions. While corrupted scientific enterprises are found all over the world, global standards of cosmopolitan science demand that technologies of moral and scientific expertise be integrated in a reputable enterprise.

Ethics of Collecting DNA

Scientists are moral practitioners in that they are morally implicated in the ways they treat human and nonhuman beings and their body substances. Science is a reflexive form of knowledge, and it is always in revision and subject to reinterpretation as new observations emerge and new answers are needed in new environments of politics and research. Thus, transgressive ethics and a form of productive uncertainty obtain when scientists must negotiate moral dilemmas for where and when conventional ethical mechanisms are not sufficient or defensible in particular research contexts. The collecting of specimens from aboriginal peoples is illuminating of whether bioethics can add ethical value to the DNA and to their donor communities.

In recent years, researchers working with aboriginal peoples, acutely aware of the inadequacy of established bioethics, have devised new practices to create "ethical biovalues." Emma Kowal reports that "ethical" researchers, mindful of decades of violence against Australian aboriginal peoples, have built new protections for their previously accumulated indigenous DNA. As guardians of indigenous DNA, they invest time, labor, and affect to "protect a *vulnerable indigenous collective body* from scientific intervention." In addition, efforts have been made to connect indigenous people to the samples of ancestors, thus rescuing stored DNA from oblivion and making them culturally meaningful to present-day aboriginal peoples. This example of ethical intervention thus enacts a "racialized biosociality of vulnerability" that is expressed through the adoption of "orphaned DNA."[15] This case of a long-delayed enactment of reciprocity reflects researchers' ideas of what is deemed "ethical" in the exchanges, but perhaps, not from the perspective of the donor community.

Here I consider how Asian researchers consider the ethical contingency of the contemporary collection of DNA from indigenous peoples, raising doubts about international moral norms as too narrowly focused on the subject's intentional decision as expressed in voluntary consent and nonmaterial exchange. A British foundation for population research identifies "universalisable moral norms" of a "moral DNA": autonomy, noninjury, beneficence, and justice.[16] In practice, while informed consent forms are assumed to safeguard individual decision making to realize noncoercive and voluntary conditions under which human biospecimens may be "gifted" for research purposes, they often come to stir unexpected ethical quandaries.[17] For instance, informed consent encodes the individual and her agency as the ethical unit to be addressed. Indeed, Asian geneticists in the field often find the focus on individual autonomy not socially meaningful to indigenous donor communities. That is, the identification of ethical practice with individualized "rights" leaves many questions of collective well-being unaddressed and tends to frame the ethical relationship as a consumer relationship with two figures in a transaction. Even the ideal of the "gift economy" often invoked to underscore the purportedly noncommercial nature of health research can appear to indigenous peoples as unethical because of non-socially and materially meaningful exchanges. While geneticists in Southeast Asia abide by international bioethical principles, they are often rankled by the inadequacy of such concepts to reflect the social expectations of indigenous peoples drawn into DNA surveys. The doubts of geneticists on the ground thus raise questions about what kinds of ethical values would be considered fairer in the calculus of donor communities.

ADDING ETHICAL VALUE TO DNA

Fresh out of Oxford, Dr. Yang led a team to teach local researchers to gather human samples in seventeen African nations. Yang, a rising star in Singapore's genomic science, hoped to help lay the foundation for genomic science in Africa, the continent with the greatest range of human genetic variations.[18] The additional goal was to help African researchers develop the capabilities to study and control knowledge of their exceptional genetic heritage.

In Africa, Yang remembered encountering many peoples making a bare living off the land. Yet their subsistence livelihood was not the first obstacle to collecting DNA. He candidly noted that it was "always a challenge to collect blood samples from indigenous populations. Most native people do not know what genes are," and so explaining the potential values or benefits of giving blood was a difficult task. He was quick to add that the same can be said of most residents of cities anywhere in the world. Like the anthropologist in the bush, Yang encounters the old questions of translation, beliefs, language, and knowledge economies when it comes to how to teach the natives to boil water. In such an encounter of cultural blindness, the implementation of bioethical norms only raised an ongoing question of whether such a measure of "ethical" behavior could be sufficient for evaluating biomedical practices in the very common situation where researchers were meeting with donor groups consenting to an experiment they may not understand. For Yang, while translating the scientific intention and goal to potential donors of blood and saliva samples, the limits of "informed consent" became clear very quickly. Because of language problems as well, Yang relied on African field assistants—often speaking languages other than the native language of the potential donor group in question—to interpret what genes and the research meant to group leaders who often give consent on behalf of their people. He had many questions. How informed should donors be, especially when the concept "gene" has no direct translation and is glossed, perhaps by a metaphor in a native or the dominant language of the country? Were the leaders expected to understand the basics of genomic research and expected outcomes in scientific or metaphorical terms?

Indeed, perhaps even most scientists and anthropologists do not "know" what genes are because they are mobile epistemic objects, or a heuristic device that already has yielded to new registers and ways of registering and testing life.[19] The act of translation would depend on a semantic stabilization of these flexible and polysemic heuristics. Given this productive indeterminacy in the gene as an experimental and "epistemic object,"[20] how could a model of "informed consent" that hinged, in principle, on the self-knowing choice of individuals deal with these slippages of translation, of concept, of register? In other words, by the nature and pragmatics of translating experiments to potential donor populations in fulfillment of the formal demands of informed consent, there was no getting away from using symbolic idiomatic language to describe the experiment or to elicit consent from would-be donors.

In rural Africa, Yang noticed that the informed consent, when given, is of a collective, not individual, nature, as intended by European standards. Because of what he called "tribal organization," Yang and his assistants made initial contact with indigenous leaders, who would then avail their tribes to researchers, in order to obtain a donor group of suitable size. What degree of voluntarism or free will can be exercised by would-be donors embedded in unequal relationships between principal researchers, local interpreters, indigenous leaders, and ordinary people? What political or cultural benefits may be at work for group leaders to agree to permit blood sampling of "their people"? Yang understood that informed consent cannot capture the actual social and subtle pressures group leaders apply to their followers. Based on his field experiences, Yang wondered whether the notion of "individual autonomy" underpinning informed consent was politically enforceable. When it comes to some non-Western contexts, should the focus of informed consent be on protection of human rather than individual free will?

At the same time, what are the nebulous implications of "giving consent" when it involves the giving of a bodily substance loaded with biological and cultural values that are central to life itself? In a telling example, Yang noted that collective permission given by leaders often did not allay related fears of indigenous peoples that local interpreters could explain away. For example, some potential donors associated the effect of needle pricks to draw blood with that of mosquito bites and worried about possible links between the drawing of blood and exposure to malaria. Here is an instance of what

Michael Taussig would call the "phantom-objectivity of disease" promoted by biomedical diagnosis, a reification that disguises or denies the situated social construction of illnesses.[21] While such cultural fears may be deemed scientifically erroneous, informed consent could not easily brush them away for individuals corralled by both their leaders and the new science to give bodily substance, for something they could not quite grasp would not endanger them. This interaction recalls Clifford Geertz's notion of "witchcraft-as-commonsense," that is, as an elaboration and defense of colloquial reason that allays the fear one endures with other affects when one faces uncertainty.[22]

I asked Yang about whether his African blood donors were compensated for giving blood. He put his finger on a conundrum by pointing to the international bioethical rule that a voluntary donation is incompatible with economic payment. This ethical standard seemed particularly unjust when blood is collected from economically needy people. "In Africa, we were not allowed to compensate blood donors. For the people, benefits may be a survival issue. Thus if we offer compensation, people may sign the form in order to get the money. One cannot then say this is a truly voluntary donation." In other words, ethical guidelines ban payment as a commercialization of what ought to be an "ethical" exchange, constituting the circulation of money as an automatic negation of the ethical status of research. In order to sustain the ethos of gift giving in DNA collection, poor people were not permitted economic rewards they may welcome or need.

Needless to say, Yang was confounded by the inadequacy of ethical norms established in liberal societies that functioned as a universal standard to minimally protect or satisfy donor groups. He sought to add symbolic value to the DNA extracted from donors by telling moral stories that could symbolically uplift research subjects, precisely by reinvesting their sense of belonging to and existing as an ethnic group. "We tell them it is good for your population and subsequent generations if you give samples for study." Such stories have the effect, he hoped, of enhancing the value of their ethnicity as a people. By casting DNA as a source of hidden stories, Yang transformed DNA into an ethical resource (or "ethical biovalue"[23]) that permits indigenous peoples new ways to connect with their bloodlines and to reimagine their past and future. Beyond the ethical goal of instilling "genetic pride," there is the larger hope that his experiment would help some Africans to map their genetic variations and would make their collective DNA data meaningful to people within these seventeen African nations where the project was sampled.

Thus, in Yang's formulation, it is the figuration of the gift rather than the logic of the gift that is linked to a mode of evaluation and standing such that hierarchy is part and parcel of the operation. When gift exchanges are conducted outside the language of political inequalities that are to be rectified by ethical regimes, then the impossibility of the logic of the gift is constantly invoked through these figures of gift by the scientists and their advocates. The next case illuminates how DNA collection can add ethical value not only to DNA, but also for indigenous peoples themselves.

ADDING VALUE TO NATIVE DONORS

Dr. Mega, a member of the Malaysian group in the Pan-Asian SNP Consortium, was in charge of collecting blood samples from the native Orang Asli (Malay: original people) in Malaysia. In the regional effort, each participating nation determines its own scientific jurisdiction, with each researcher collecting biospecimens under the institutional approval of his or her country. But even within the same nation, different practices were deployed. A policy reviewer of the SNP Consortium, Dr. Lee, reported that in Malaysian Borneo, an initial way of getting informed consent to collect DNA from the Dayak was through a "video talk to the headman," followed by each individual donor providing a "thumb print" indicating consent (sometimes also videotaped). In addition, "usually, the respondent gives a verbal account of ancestry, parents and grandparents' ethnicity." By contrast, Mega, who was conducting research among the Negritos in Northern Peninsular Malaysia, discussed an elaborate set of caring practices to help sustain marginalized indigenous communities.

From the start, Mega decided, "The Orang Asli are not there to be exploited," and the experts carefully managed the delicate process of contact and building of trust. In her initial foray, Mega led an interdisciplinary team with expertise on genetics and infectious diseases to contact a small Negrito hunter-gatherer group living in the jungle. They began with a first snapshot of the health status of this Orang Asli population and asked them what they needed. In response to requests for material goods (rice, salt, clothes, toys, and books), the team returned lugging rice bags and other items for the two-hour upriver trip. Because the research was anticipated to be long-term, such supplies, together with the provision of health services, arrived on a regular basis. Mega's team in effect adopted the Orang Asli group even before they began gathering samples, and they have had an ongoing support relationship ever since.

"Ethics and norms," Mega asserted, "should be relevant to community, not transposing norms from elsewhere." The Pan-Asian SNP Consortium follows a combination of international ethical standards, local internal review board (IRB) standards, as well as "local informed consent norms" in the recruitment of all subjects and handling of samples. Each of the fourteen participating countries must submit copies of their IRB approval to the consortium's policy review and ethics board. Mega's own study on single nucleotide polymorphisms (SNPs) was supervised by review boards in the Malaysian Ministry of Health and by Melbourne University, her employer. Whereas the Malaysian authorities were concerned about obtaining health information, the university review board encouraged her to interpret ethical guidelines in context.

Mega said, "Informed consent is what you interpret it to be. Get across the [ethical practice] as best you can. What cultural norms are locally [relevant], and how do they synergize with the consent documents?" She insisted that the Orang Asli identified with and agreed to be part of the study from the start. With social trust solidified by material supplies, the Orang Asli agreed to provide DNA specimens, depending on their comfort level, from their blood or saliva samples. Despite ethical conventions banning nonmaterial exchange, for Mega's team, the material sustainability of indigenous peoples trumped fears that the DNA may appear to be commodified, made implicitly obligatory in the relationship that the researchers cultivated with the tribe.

That ethics included the custodianship of the donor community, a guardianship that necessarily included efforts to make population genetics meaningful to the Orang Asli. Their DNA "tells them where their ancestors came from. Would they join us on a journey of discovery?" Mega enacts an old practice of the anthropological search for origins by transposing onto the "closest" living native some imagined "originary" ancestor. Her genetic analysis of mitochondrial and SNP data allows Mega to link biology, motherhood, and identity. "The story of your ancestors [orang dulu dulu] can be traced through the mother's line. Your mother's body bears secrets to your past and future." Her ethical attempts to add value to the DNA went beyond a onetime storytelling. On Orang Asli Celebration Day each year, the team returned to the settlement to reiterate the social significance of the DNA findings to "where your people came from." Researchers showed the Orang Asli within a map of early human migrations out of Africa. The Orang Asli liked not only stories of origins but also the fact that their health was taken care of by a DNA team linked to the national governance of indigenous peoples. By alternating between adding value to the collected DNA and to their corporeal owners, Yang and Mega tried to recognize and integrate what would be considered ethical from the perspective of donor populations.

Building a Biomedical Platform

A one-sided picture of scientific entrepreneurialism in emerging countries suggests the routine avoidance of ethical norms in experiments wherever possible. The implication is that in new biomedical centers, adherence to ethical standards is merely a strategy to reel in bioprofits down the line. At Biopolis, there is serious implementation of ethical "best practices," but ethical norms are but a necessary first step in a rigorous, costly, and uncertain process of entrepreneurial investments necessary for testing and making novel drugs that are marketable.

According to researchers I interviewed at Biopolis, ethical norms are only one element in an optimal set of conditions—scientific, economic, cultural, and social networks—that shapes entrepreneurialism in the life sciences. As smaller pharmaceutical companies search the world for ideal locations for drug-testing operations, the real stakes are reputable centers of scientific excellence and innovation, new markets, and the production of globally marketable drugs. For globally tradable brands, ethics is not an impediment to drug development but rather its condition of possibility. However, what that means, as I have been saying throughout, is not always the right ethics for the specific type of science in context.

WHY SINGAPORE?

As global companies seek to test drugs outside the West, it is not the variability but the uniform application of informed consent in multiple sites that creates synergy across clinical labor markets.[24] Medical experiments are costly ventures in part because of the time it takes to develop new drugs and the uncertainty of their effectiveness and safety for participants and potential patients. Scientific and ethical regulations dog every stage of clinical trials, so each experiment runs a gauntlet of stringent health and research review. Instead of being absent or variable, ethical procedures are vital operations for achieving scientific and pharmaceutical end goals.

To reiterate, Singapore models its biomedical work off the United States, well represented by the Duke-NUS Graduate Medical School. Johns Hopkins Singapore (JHS) is the country's first medical facility that combines clinical research and teaching with patient care services in cancers most common among Asian patients. In addition, there is the Singapore-MIT Alliance for

Research and Technology (or SMART) and the Stanford Singapore Bio De-sign, both on the NUS campus. U.S. medical schools are seeking the expansion of international care, shifting their center of gravity from the United States and Europe to Asia. For them, Singapore is the gateway to all of Asia, permitting them to develop well-regulated programs there that are difficult to do from afar in the United States.

Hearing about my research, a group of NUS postdoctoral researchers invited me to dinner. Dr. Sommers, a Berkeley PhD and postdoctoral fellow at SMART, gave a keyhole view of why Singapore is such an important location for Western biotech interests. For global companies, he said, "Singapore makes it easier to get in here than China." His friend from Taiwan said, "Many companies will not put their IP [intellectual property] crown jewels in China. . . . China wants to grab technology from overseas without paying what it calls a technology tax." By comparison, Singapore is an established headquarters for multinational firms that seek links to universities and good science workers who are cheaper than in the West. Sommers summed it up by saying that "Singapore is a safe haven where everything works, top-notch education, and as long as the infrastructure is good, scientists will come here. Drug companies are attracted by Asian samples, patients, and markets."

When it comes to Asia, I was told, global pharmaceutical firms find Singapore a safe bet, because the city-state carries all the operating costs and is a signatory to all IP treaties, in line with American and British practices covering the making and marketing of drugs. IP protections ensure that patents turned into products can flow back to the national community. For instance, the mission of JHS is to expand beyond the United States, by looking at genes specific to diseases found more often in Asia. The goal is to develop patents for licensing out or to form a startup company (in which JHS will own a piece) and thus draw royalties from the final products. The expectation is that Singapore will become a site for the listing of biotech firms in Asia and for attracting private venture capitalists. Singaporean firms such as Temasek invest in opportunities in the region. In contrast, South Korea and Japan, partly because of their more nationalist framing of science, have made few attempts to invest in biotechnology industries outside their countries.

Whereas other experimental sites in Asia may recruit offshore proxies (for American patients), Singapore has in place regulatory and technical regimes for accessing and testing research subjects for new drugs that are customized to Asian populations. Informed consent is an important mechanism for delivering volunteers and patients to participate in usually Phase I trials, and

subjects receive "good reimbursement" for travel and time lost. The potentialities of developing customized medicine for new Asian markets demand
the careful selection of subjects and prevalent deadly diseases that all come to
bear the indicators of Asian ethnicities. Thus, informed consent is an enabling
standard practice that supports the clinical testing of customized drugs.

Drug companies cannot risk bad reviews, and the focus is on high-quality
research that produces new therapies that can win approval as global commodities. There are therefore no "easy passes" when it comes to ethical norms, even
in developing countries, but rather very high hurdles to clear. But international
bioethics is merely one form of best practice among many that global drug
companies seek in biomedical platforms. High standards of data management, quality controls, research organizations, legal protections for IP, and
cosmopolitan business practices are all needed in order to draw clinical trials
and sustain the reputation of new biomedical centers.

THE "ONE-STEP SOLUTION"

Cancer may be "the emperor of all maladies,"[25] but it is also a variable disease demanding new empires of investigation. As discussed earlier, Singapore
has rapidly risen to the challenge, seeking to be "a one-step solution" for testing
new cancer drugs in Asia. Pharmaceutical giants (Pfizer, Eli Lilly, and Merck,
among others) signaled the future of cancer medicine when they recently
formed an "Asian cancer research group" in recognition that there was an unknown continent of cancer subtypes for which to develop new cancer diagnostics and treatment.

Singaporean science strategy was to build a genetic cancer database so that
researchers could hunt for ethnic signatures in the diseases. In the words of a
local physician, the accumulation of ethnic biomarkers for cancer "can make
companies rich and encourage researchers to develop more such tools."[26]
Singapore's strategy to be the cancer capital of Asia is often expressed in the
business school idiom of Harvard University, which many have trained at or
visited. Scientific entrepreneurialism is both practice and pitch in that ethical norms must be augmented by "quality metrics" of up-to-date systems and
cultural skills to deal with politically sensitive Asian neighbors drawn into the
network of biomedical experiments.

Singapore's rising empire of cancer is centered on Biopolis and Hospital
Hill, where Duke-NUS is located amid a cluster of public hospitals and cancer
centers (refer back to figure I.2). Dr. Loh, an ebullient "head-neck-and-throat"
oncologist, was the head of the National Cancer Center. His cancer building

overlooking the island was at the nexus of three major entities—digital medical records, a tissue depository, and a single review board. He explained that
by linking the critical hospital and research institutions in a chain, Singapore
operates "like a biobank" in bringing together health data and samples. Other
cross-linkages to units such as immunology, immunotherapy, microorgan-
isms, infection, and transplant facilities also enhance the streamlining of research access to data, tissues, and review procedures.

I asked about the place of ethical norms in this cancer research infrastructure. Loh reassured me, "We have standard operating procedures [sops] for
good health-related research." Ethical best practices govern the collection and
management of data and tissues for research purposes. Patients sign consent
forms that transfer ownership of their cancerous tumors to the center, which
makes their medical records immediately available to researchers as anonymized data. The centralization of digital medical records supplies anonymized
data, tracking patients from birth to death, which is then fed into a national
disease registry. In addition, there is a national cancer repository linked to the
medical data. "The whole of Singapore is aligned," Loh insisted, where ethics
are concerned.

Besides the ethical handling of patients, data, and samples, ethical reviews
are an integral part of the digital-bureaucratic system of managing and channeling data flows. Ethical and research protocols are coordinated in a way that
makes it easy to conduct reputable experiments. Reciprocal exchanges link
all IRBs in the nation's hospitals, clinics, and labs so that there is a "one-stop
process of gaining access to data." Because of the unbroken chain for the data,
samples, and review process, Loh boasted that "when it comes to clinical trials
for cancer, we are the data managers" for Asia.

Indeed, Dr. Loh implied that the high ethical standards combined with
great infrastructure more than make up for the city-state's small population
(5.5 million, half of them citizens) when it comes to testing drugs. Singapore specializes in Phases I and II of clinical trials (a view endorsed by a
McKinsey report). Phase I involves about twenty human volunteers, both
healthy and sick subjects. Phase II involves fifty patients with the disease to
be tested with selected drugs. The goal is to see positive results: Does the
tumor regress? Does the patient live longer? By Phase III all subjects are patients. It is a randomized process in selecting patients with the same stage
of disease to see if the treatment becomes promising. This phase of testing
needs thousands of research subjects but despite the island's small popula-

tion, Loh claimed that, in Asia, Singapore has "become the thought leaders in coordinating Phase III trials." Despite the perspicuity of their strategy, Loh and his colleagues are implementing an engineering-technocratic logic of philanthropic commerce.

Many companies turn to South Asia for Phase III drug tests, but Loh noted that "there is a common problem regarding quality of standards." Here Singaporean scientists step in as ethical middlemen armed with a degree of "cultural sensitivity" to enforce ethical practices overseas. I asked him how this could be done. "We need a ground-level approach to see how questions are worked out, to enforce trial monitoring, to treat our partners as equal. We subject trials to three international reviews that oversee the data and can stop trials at any point if things are ethically in question." In addition, Loh said, every six months, Singaporean researchers did random checks on data follow-up in overseas trials. They impressed on Indian partners that only those experimental subjects who were allowed regular follow-ups on their health could be used to capture information. He mentioned that recently a clinic in Pakistan had been dropped from a trial after it had three failures. Loh's claim is that Singaporean scientists act as "trial monitors" by participating and spreading ethic practices (at the "SOP passing level") not only on the island but elsewhere in Asia.

What did Loh mean by "cultural sensitivity" in carrying out experiments in different Asian spots? In what sounds increasingly like a sales pitch, he noted that, as a tiny nation, Singapore once again has an edge. Its multicultural researchers were acutely aware about Asian countries' sensitivity to their relative geopolitical status and ranking in a time of rising scientific competition. Thus, he said, clinical trials led by PRC Chinese or Indian scientists may not be "well accepted" by their big power counterparts. "It is a case of elephants fighting over the turf." More broadly, Loh mentioned that American scientific brashness often offended Asians, thus implying they needed Singaporean mediation to smooth ruffled feathers. He cited an example of Western companies that used avian flu samples produced in Indonesia and then sold expensive vaccines back to the country.[27] Another case of cross-cultural insensitivity was a U.S. company that, having sequenced the genome of basmati rice, asked Indians to pay for the work. By contrast, Loh added, "Singapore is one of the better places to run clinical trials, provided we are not patronizing," a moral quality that Singapore too may be susceptible to as Asia's rising star in cancer research.

The challenges and uncertain outcomes of clinical trials, therefore, demand a kind of scientific entrepreneurialism that combines ethical norms in an intricate mix of scientific and cultural practices to ensure optimal conditions for successful experiments. Finally we see how another Singaporean oncologist juggles many tasks, including cultural sensitivity to big pharma.

Asia has become the ground zero of clinical trials because of the changing dynamics of global drug markets. Meanwhile, big pharma firms, having steadily scaled back costly in-house experiments, are looking for promising lines of research by scientists in research universities and hospitals within the region. In Singapore, generous funding of "investigator-initiated trials" is a strategy to suck in investments by drug corporations. The goal is for the city-state to become "a healthy pipeline for new drugs," especially for "Asian cancers." "There is an art to producing vaccines," Tan said. Trained as a human stem cell biologist in the United States, while working on colon cancer, Tan was among the first researchers in Singapore to develop cancer vaccines for human beings. As he talked about his experimental trials, it seemed clear that the art was expressed less in fiddling with cells than in managing the cultural perceptions of big U.S. companies.

In Singapore's research labs, one often encounters an atmosphere of American scientific ebullience and Asian cultural particularism. Many Asian scientists had either been trained in universities in the United States or had spent research time there. There was easy use of American military jargon to describe diseases and therapies. Emily Martin has observed the use of military images in American models of the human immune system where specialized white cells are compared to an internal arsenal that can identify and attack invading "enemies" or pathogens.[28] In Singaporean labs, similar images are deployed in describing the role of immunology in Asian cancers.

Describing his new immune therapy for nasopharyngeal carcinoma (NPC), Tan mixed U.S. military metaphors and Asian ethnic differences. Many types of cancers and cancer-related diseases among Asians were due to auto-immunology, or "a case of friendly fire," where the immune system as an armed force fails to recognize its own. Tan was excited about his immune experiment because it enlisted the body's internal defense to fight solid tumors without hurting the patient.[29] Comparing himself to a commando, he said he "trained T-cells "(white blood cells that mediate immunity) from the patient to deliver antibodies to cancerous cells while sparing healthy ones.

This method skillfully avoided friendly fire while using the body's own cells to build up the body's own long-term resistance.

Tan then gave a familiar sales pitch on Singapore's ethnically matched biomarkers. "When it comes to Asian cancers, we have a handful we can be competitive with: NPC, gastric, breast, lung, and cell lymphoma. We have an advantage in multiracial groups, that is, ethnic groups that metabolize drugs differently." It seemed that Tan had made these remarks many times before, to U.S. drug companies, because it was clear that Singapore's ethical best practices and great infrastructure were not enough. Drug companies looking for promising new therapies would expect "social and cultural information" to design personalized medicine. Tan implied that it was up to Asian researchers to define what sociocultural data would be and then make it operationalizable for drug testing and customization. Thus, the capacity of researchers to convince big pharma that their data and experiments capture specific genetic, molecular, and behavioral features is a strategic selling point.

In the middle of our interview, Tan took a phone call to plan a restaurant date that evening with a representative from a big drug company. He made a joke that inviting drug company officials to a Chinese banquet might make them want to hear more about his many initiatives for cancer treatment. Putting down the phone, Tan switched to a more somber tone as he explained that a scientist needed to woo big pharma in different ways. The suggestion was that the ethnic sell had to be augmented by familiarity with American culture. Knowing that I was from the San Francisco Bay Area, he compared his highly skilled program to the running of French Laundry, a three-star gourmet restaurant in Napa Valley, consistently rated among the best in the United States. In order to run his immunology project, he operated like "the restaurant manager who has a top-of-the-line kitchen and needs to hire chefs, train them, and buy lots of ingredients that are not readily available or packaged." Contrast this with the sentiment in his opening quote, when he suggested that scientific entrepreneurialism is like running a franchise for Starbucks.

In other words, he had to assemble various ethical, ethnic, and professional elements in order to run his lab and design an experimental trial that might ensnare American business. "We have U.S.-linked research but with an Asian flavor. The hope is that companies will think that we can work as well as scientists in the States and that we can be the launch pad for new drugs and therapies." He also emphasized Singapore's extensive collaborative networks: "We can link up with other scientists in their specialty all over the world. For example, for lymphoma problems, we can call up experts at Stanford Medical School."

In this Asian scientific entrepreneurialism, cross-cultural skills, more than ethics, may be a way to play to U.S. corporate expectations. For an emerging biomedical frontier, scientists need forms of social as well as ethical capital, not to mention sheer expertise. Besides demonstrating familiarity with U.S. corporate culture by making knowing references, they pitch Asian ethnicity both as a site of ready-made expertise as well as a horizon of coming market possibilities.

Above, I illuminated how managing the ethical and regulatory gaps between researchers and human subjects is a critical part of the science profession worldwide. The productive uncertainty of bioethics is registered not by its absence but by its provisional capacity to satisfy the moral, social, and regulatory conditions for reputable pharmaceutical experiments. The uncertainty of bioethics as practical reason is therefore productive of additional ethical and regulatory interventions in particular contexts of experimentation. First, conventional, universalized ethical norms are revealed to be provisional and limited in meeting social obligations to research subjects in many places. In some countries, cultural elders and/or the state define the horizon of the common good for the donor community, therefore requiring researchers to enact additional ethical protections besides those indicated by international bioethical guidelines. As situated moral practitioners, researchers often find gaps between global bioethical norms and community values on the ground, and they must engage social expectations of the donor community in order to make the relationship a more satisfying one for the latter.

Second, the uncertain efficacy of bioethical guidelines opens up another gap: one between modern biotechnologies and what modern cosmopolitan science is, or ought to be. Biotechnologies have proliferated around the world, and yet, in too many places, there are no coherent conditions that regulate their proper uses and applications. Reputable science requires that bioethical norms and biotechnologies be governed by regulatory and technical regimes that ensure consistency in accessing, treating, experimenting on, and testing of human (and animal) subjects. As some scientists cited above were eager to claim, Singapore is an exceptional regulatory and cultural zone within which cosmopolitan science in Asia can be safely conducted.

FIVE

VIRTUE AND EXPATRIATE SCIENTISTS

> Biology knows no borders. . . . Scientists have
> become very mobile, attracted by funding,
> infrastructure, equipment, and new bioresources.
> —DR. MA, Genome Institute of Singapore

Situated Virtue

In much of contemporary Asia, medical science is valorized as the highest form of modern knowledge that can cure national backwardness and promise a path to a flourishing future. Medical doctors were among the most outspoken of anti-imperialist nationalist fighters across Asia. A major figure was Dr. Sun Yat-Sen, the nationalist founder of early modern China who sought to "heal the nation" ravaged by opium and other calamities introduced by foreign imperialists. Other examples include Dr. José Rizal, the revolutionary hero of the Philippines, and more recent leaders of independent nations.[1] Colonial Western views of poor Asian countries rendered them as places of illness, as in representations of early twentieth-century China as "the sick man of Asia," as well as of the Southeast Asian tropics as rife with florid tropical diseases, thereby causing natives to be lethargic and thus economically stunted.[2] This colonial association of Asia with pestilence, indolence, and backwardness infused modern medicine in postcolonial times with an indelible political dimension. Today, most leaders in independent nations embrace the role of modern science in saving lives, improving productivity, and even advancing democracy.[3] But as Asian "tiger countries" leap ahead in economic competition, the mastery of biomedical sciences perhaps presents more daunting challenges.

Tiny Singapore has had its own political dramas with modern medicine. From the colonial-era initiation of the fight against parasitic diseases, to bringing the severe acute respiratory syndrome (SARS) epidemic to an end in the late 1990s, medical expertise has always been closely tied to public interests and national survival. This intertwining of virtues in civic duty and in science as an aspect of nation building became more complicated when the city-state reinvented itself as a regional biomedical hub at the turn of the century. The Biopolis milieu therefore presents a fascinating laboratory for exploring how, as cosmopolitan science migrates across the globe, virtue in science becomes complicated, especially when expatriate professionals may be driven more by professional incentives and career pathways than by a personal duty to serve the common good.

Writing about the evolution of twentieth-century American science, Steven Shapin traces the growing tension between "science for its own sake" and a job increasingly burdened with practical ends.[4] He invokes Max Weber's notion of "science as a vocation" in a sense akin to a religious calling.[5] In postwar U.S. and European science institutions, he argues, this notion of personal virtue combined with public interest underpinned the professional standards and affective relations in what have come to define cosmopolitan science. Virtue in cosmopolitan science also suggests a willingness to persist in painstaking research and in obscurity, sometimes for decades without ever achieving recognition or success.

This persistent virtue in North Atlantic science worlds, which led to the discoveries of DNA and RNA, has increasingly been undermined by the very biotechnologies it created. In what has become an exceedingly competitive industry, especially in the United States, scientists are increasingly driven to convert their discoveries into biotech startups that can earn millions (the reigning example being J. Craig Venter). Therefore, Shapin argues, in the midst of such radical uncertainties, the personal virtues combined with the charismatic authority of principal investigators (PIs) are crucial to the advancement of "science that matters."[6] There is anxiety that commercially driven bioscience will dissolve ethical restraints regarding experiments that may be potentially lucrative yet harmful to life on earth. Sydney Brenner, a pioneer in genetic biology and an advisor to the Biopolis endeavor, had cautioned that a "scientifically responsible ethics" requires us to "distinguish very carefully between impotence and chastity; it is important to distinguish between 'can't' and 'won't.'"[7] Especially in American science, some worry that virtue (doing science for its own sake but tethered to public interests) has been hijacked by capitalism in the figure of a relentlessly expanding pharmaceutical industry.

There is the question of whether the conception of scientific virtue can be applicable in contemporary times or can universally be the same. Indeed, Max Weber wrote in 1917 that the nineteenth-century ethos of "science as vocation" was already not tenable and sustainable in the early twentieth century.[8] His modality of disinterested science is not remediable through what is taking place in the twenty-first century. In a related focus on a general moral theory, Alasdair MacIntyre bemoans the fragmentation and lack of coherence in the moral countenance on virtue he traces to classical Greece.[9] Here, anthropologists are well placed to challenge MacIntyre's diagnosis of malaise in ethical discourse by offering an alternate analytics of contemporary ethics as context-dependent. Stephen J. Collier and Andrew Lakoff argue that, in each political-cultural context, situated interactions of "technological reason and biopolitics" are sources of dynamism critical to questions of "how to live" and practices that shape the constitution of ethical subjects. Instead of individuals reflecting on virtue, ethical norms in formation involve communities of experts adjudicating among values, concerned not only with "the life of citizens but of biological and social beings."[10] In global times, then, technology, biopolitics, and ethics generate "regimes of living" that flexibly address what is at stake for the common good in that particular situation.

We can now return to the question of what happens to virtue as configured by postwar American cosmopolitan science when it flows to other political and normative contexts of experimentation. Clearly, virtue, scientific and otherwise, has to be examined in contemporary science milieus (each one a specific assemblage of state-capitalist interests, interdisciplinary expertise, and transnational resources) and the situated modulation of the meaning of doing ethical and responsible science. Scientific virtue comes in different forms—as love of science, serving public interests, humanity at large—but also perhaps taking risks while ensuring that good science is being propagated across the world.

In the developing world, the translation of virtue would be that science (usually a state sector) has the primary obligation to solve problems of the health of the nation. Science as an institution therefore has a kind of moral authority that exceeds the personal virtues of individual scientists, the focus of Shapin's argument. In recent decades, this orientation of science toward public interests includes developing the biomedical capacity to respond to looming problems of biosecurity. In Singapore's case, the state configures the normative space in which the achievements of individual PIs are consistently affirmed not as a matter of simply individual genius, but also as moments for the accrual of national and "Asian" prestige. This ethos is different from but nonetheless reminiscent of postwar Cold War military science in the United States.[11]

In Biopolis's state-supported milieu, scientific virtue and authority have to manifest commitments to the island's population or the region at large. Top scientists such as Edison Liu strove to make visible its capacity to do public good. Many of his over-the-top pronouncements can be interpreted as the need to articulate virtue amid a global expansion of the pharmaceutical industry. As a Hong Kong–born American oncologist, and the first science leader and spokesman for the Biopolis initiative, he had a special pulpit from which to declare the biomedical science as a necessary public good. Over the first decade, his "charismatic authority" geared toward institution building[12] and demanded a routine performance of the legitimacy of a state-supported biomedical enterprise that is not entirely commercially driven, nor neglectful of practices of living with respect to the good. Here is an alignment of personal virtue and contemporary ethics in that bioscience in Singapore should not just be productive and successful, but attentive to the biological and social stakes of life and living in the Asian tropics. Liu personified Biopolis as an enterprise that is as much about building a regional biotech commons as about furthering biocapitalism—in Asia. Therefore, virtue becomes part of a negotiation between two versions of the nation-state, the socialist-nationalist Singapore of the past, versus the cosmopolitan and regionally oriented one, built out of incentivized foreign talent and oriented to a region far beyond Singapore's shores. But in late 2010, the Singapore Ministry of Trade, overruling the Agency for Science, Technology and Research (A*STAR), decided that, going forward, Biopolis projects had to be based more on partnerships with private entities. This forced march toward an industry-aligned mode was devastating to Liu.

During his decade-long tenure as director of the Genome Institute, he tried to balance state paymasters' demand for short-term metrics with his larger vision of Biopolis as a center for confronting emerging biothreats in Asia that also menace the world. Having lost the battle for advancing science for the collective good, Liu resigned and returned to the United States.[13] Nevertheless, Liu's vision of a Biopolis for Asia left its stamp on the institution, state funding has continued (though at a reduced rate), and Biopolis is now a successful milieu for conducting cosmopolitan bioscience outside Euro-American settings. In his model of entrepreneurialism, a "scientifically responsible ethics" is actualized in the work of practical institutional building, binding the bioeconomy to biosecurity capacities in a region that needed them. As state funding steadily gives way to private investing, a "pure" notion of scientific virtue unadulterated by market or public/national agendas is unsustainable.

The variable and contingent ways that virtue in science can be shaped and expressed can be tracked in the stories of expatriate scientists who represent a major component of the research milieu in Singapore. Biopolis is exposed to normative uncertainty of a greater variety, for it is physically situated in Singapore but largely sustained by an international talent pool drawn from across Asia, Europe, and America. Here we take a closer look at how virtue is somewhat contingently influenced by the pursuit of a science career in the vortex of mobile science, global capital, and situated politics.

Superstar Scientists

Noting that Biopolis has emerged in a period of "fast science," by which he means rapidly evolving science, Dr. Ma, a leading stem cell PI, said, "A project like this needs long-term state commitment, including the decision to get highly skilled expats" and to pay them at competitive global market rates. The life sciences have become globalized, and at Biopolis, this globalization has generated the growth of infrastructure and the collection of scientists. As mentioned earlier, Singapore researchers work closely with colleagues not only in the United States, but also in Europe and the Asia-Pacific in developing research projects and PhD training programs.[14] At the same time, Biopolis draws ambitious researchers from many sites, in part because "world-class" scientists and their junior PhD workers consider moving from one prestigious lab to another in the course of their careers. Singapore has become a global pied-à-terre for mobile scientists.[15]

This chapter shows that scientists at Biopolis come to the issue of virtue in different ways. For the many scientists routing their career paths through the island-state, personal and career interests are predominant. We hear their expressions of self-interest and altruism (if any) and the benefits of living in Singapore. In contrast to the expat PIs, young lab technicians recruited from

India and China tend to view science mainly as a job that enables transnational relocation from their crowded homelands. Both categories of foreign scientists are vital for sustaining Singapore's international research milieu and running the research labs. The question is how long can the Biopolis enterprise depend on the influx of expensive foreign workers? I then turn to Singapore's efforts to "grow Asian talent" through generous scholarship programs to ensure the sustainability of the bioscience project. Here, the program comes up against a different notion of virtue shaped by Singapore's demographic limits and of their cultural "fear of losing out" (*kiasu*). I conclude by suggesting that fear of failure is also a virtue in that it prods risk-taking and grappling with uncertainty surrounding Biopolis's future.

By the time I started my research in 2004, one top scientist, Alan Colman, had sworn off interviews, especially about stem cell research, and just wanted to do his work in peace. I was able, however, to talk to less publicly recognized but prominent scientists in their own right. The overall themes expressed by expatriate American scientists were the desire to undertake new research and build new programs without hassles and hurdles encountered at home, have access to "Asian" medical information and tissues that make for novel findings, and enjoy excellent salaries and the excitement of living in Southeast Asia. In addition, the block funding structure meant that scientists did very little proposal writing, and they get ready access through the funding system to hardworking PhDs from China and India who conduct the actual research.

"SINGAPORE IS BOUNDLESS, EXCITING, EXPANDING"

Dr. Thompkins, an American geneticist and oncologist in charge of academic training, joined the Genome Institute in 2002. He ran a postdoctoral program (affiliated with the National University of Singapore [NUS]) with twenty students. He also supervised ten postdoctoral researchers, most of them from China, in his lab. Thompkins joined Biopolis because he was disgusted with strictures on science in the United States in the years immediately following 9/11 "when venture capitalists were sitting on their wallets and the government was more interested in bombing foreign countries than in funding stem cell research." At Biopolis, Thompkins conducted stem cell research related to cancers, infectious diseases, and other work related to translation medicine.

He stated that, opposed to the American labs where he had worked before, "Singapore's approach is more academic type research in an entrepreneurial environment; it is boundless, exciting, expanding. The goal is to attract drug companies to spin out new technology, to seed new companies, to train students, and to provide an intellectual foothold in Singapore." He was a leader in a disease-driven project to track the population in Asia, which has a higher prevalence of certain diseases like liver cancer. He also indicated his interest in breast cancer, liver cancer, hepatitis B, infectious diseases, dengue fever, and pneumonia. Besides, Asian populations seem to have different responses to fatal diseases. "Infectious diseases are really hot. The question is why some [Asians] don't die, for instance, from SARS (only about 10 percent do)."

Thompkins felt accepted and welcomed and thought that Biopolis's scientist-administrators were "pleased and even grateful" that he was building something there. He enjoyed the cultural diversity of places in Asia he can visit. He was "so well paid that there is no inducement to be a PR [permanent resident]; in fact, becoming one may actually reduce my recompense" because funding packages were tied to attracting foreign talent. One aspect of his job was to train more Singaporeans and to recruit Chinese and Indian PhDs who hopefully will settle down in Singapore.

"DUKE-NUS CAN BUILD A BETTER MOUSETRAP"

Dr. Fisher was a leader in cancer and stem cell biology at the Duke-NUS Medical School. He left the Midwest for Singapore in order to grow the program in stem cell biology, which stood at over seven hundred employees in 2010. He commented that in the early years there were fractional faculty from both campuses in Singapore, but the Duke faculty has dwindled to a couple. Duke-NUS in his estimation is nonetheless a viable enterprise. The new faculty was mainly recruited from NUS. I asked why Duke was interested in this arrangement with Singapore.

Fisher replied that North Carolina was "not making a lot of money here." He cited Duke's commitment to international education in medicine, pointing to another campus it had established in Hangzhou, China. After hedging a bit, Fischer said, "The vice-chancellor of Duke Med School, Dr. R. Sanders Williams, said that there is a great match between Duke and NUS in infectious diseases, cancer, diabetes. First we need to build a presence here, to establish an American-style medical enterprise." A major benefit was that researchers worked on "Asian diseases"—lung cancer, gastric cancer, carcinoma (liver cancer)—"that are common here, and [they could] have access to patients." The school also draws on medical data, tissues, and mutations from Singapore's public hospitals on Hospital Hill that have become part of the research apparatus in the Biopolis ecosystem. Such partnerships with Asian sites allow the Duke medical school to go global and to capitalize on access to Asian disease data and patients. Such partnerships smack of medico-market imperialism, involving the latest form of technology to be sprung on the region. But Singapore is an independent country that invites the transfer of cutting-edge science expertise, using it to network and leverage a form of scientific prominence in Asia.

Duke-NUS has five programs: cancer and stem cell biology, cardiovascular and metabolic disorders, health services and systems research, neuroscience and behavioral disorders, and emerging infectious diseases (see chapter 8). Fisher's stem cell biology program focuses on cancers that affect the Asian community in general. He noted that "the best science is in the cancer field, and there are many Asian-specific forms to train the best students." He pointed out that training in neuroscience would make Singapore competitive in an important area, and there is an obvious advantage in making Asia a site for studying emerging infectious diseases. From these remarks, it was clear that Duke made a strategic decision to invest in and profit from the emerging medical tourist market by undertaking cutting-edge biomedical research in Singapore. Fisher summed it up this way: "Duke-NUS can build a better mousetrap and study diseases neglected in the West—dengue, gastric cancer—in Asia, where we have access to samples and patients."

What were the personal reasons for American scientists to work in Singapore? "American scientists," Fisher claimed, "tend to believe in the mission, here at Duke-NUS, of training physician-clinicians." In an aside, he said, "the situation in the U.S. is screwy right now; the goal of many medical students is emergency medicine" rather than doing basic research. Furthermore, Duke-NUS was the first physician-clinician medical program in Asia, and Fisher and his colleagues wanted to help build up the U.S. model that links medicine and science, in contrast to the British system, which had dominated in Singapore owing to its colonial history. Besides quipping that he wanted to get away from his mother-in-law, Fisher said his children were at the right age for him and his wife to move to Singapore. "The financial constraints here are not so high." He found living in Asia exciting, and "the people really think differently about many things." Last time I checked, Fisher was still leading the cancer program.

As expat scientists, Fisher and Thompkins suggest that the major reason for moving to Singapore is because they prefer the research milieu, remuneration, infrastructure, and living conditions. The notion of virtue comes up faintly in Fisher's mission to build up cancer research in Asia within an American medical framework. Both science expats draw sharp distinctions between being scientists in Asia and in the United States, especially when it came to generous funding, and access to "Asian" information and tissues, as well as good students, for doing novel research into diseases.

"THE U.S. NEEDED A STRONGER PRESENCE IN ASIA"

Dr. Wallace is yet another U.S. scientist who was disgruntled with research conditions at home, and he came to Singapore looking for a challenge, conducting research in the Asian tropics. As a former employee of the Centers for Disease Control and Prevention (CDC) in Atlanta, Wallace told a story of emerging infectious diseases (EIDs). After World War II, Allied forces conducted research in the Asian theater on the effects of IDs on troops. Americans had developed antibiotics and vaccines like penicillin and other treatments for vector-borne diseases and were developing a new confidence in the control of infectious diseases. In the 1950s and 1960s, partnerships between the United States and the World Health Organization eradicated yellow fever, global malaria, and, using DDT, produced major public health victories. Smallpox was soon eradicated in Latin America and Egypt.

As an epidemiologist at Johns Hopkins in the 1970s, Wallace noted that the perception was that IDs as a major problem had diminished, and research was redirected to President Nixon's "war on cancer." That decade saw the National Institutes of Health (NIH) funding "magic bullet" high-tech medicines, stressing curative and not preventive medicine. This was the "era of complacency," in the use of drugs for prevention, containment, and control.

He did not mention that Nixon also took the United States off the gold standard, precipitating a new era of virtual financial trading that permitted the kind of venture capital that after the "dot-com" boom has solidified as a modality of investiture to commercial biotech and science. The medical dreams that venture capital both engendered and enacted in the United States then went global.

Rather, Wallace focused on infectious diseases as a threat that did not go away. The world was also entering the period of the jet plane, when goods and people moved around more frequently, and of economic growth and urbanization in Southeast Asia. For health reasons, Wallace "felt that the U.S. needed a stronger presence in Asia." He moved to Hawaii, the gateway to Asia, but found the universities "didn't have enough funds and top administrators didn't own the tropical diseases program."

So when Duke University approached Wallace, he jumped at the chance to be the leader of the EID program at Duke-NUS because he saw Singapore as "the driving force in combat against IDs" because of the influence of its recent experiences fighting SARS (severe acute respiratory syndrome), the Nipah virus, and dengue fever. The short spell of immunity the island enjoyed had greatly decreased as migrants brought in thousands of viruses. He noted that in 2009, forty million people and their pathogens came in through Changi International Airport. "SARS was a wake-up call for Singapore, building awareness of incoming pathogens." Duke University saw an academic opportunity to combine lab science with clinical medicine in Asia and make an impact in the region, thus building research that contributes to public good, and expands Duke globally. He had a vision of Singapore providing leadership in building something after the CDC model, but Singapore was reluctant to have its money spent overseas.

So he enrolls the CDC, the NIH, and the Gates Foundation, along with Duke-NUS, to provide opportunities for training and collaboration across Asia. In doing so, Wallace builds on forty years of working in the region. "These collaborations are all based on trust." He procures local, clinical, and epidemiological samples regularly from Vietnam, Cambodia, and Sri Lanka, and he has worked at the border of Vietnam and South China with PRC scientists on the Yunnan-Vietnam-Guangdong nexus on encephalitis, as well as neurological and respiratory diseases. At the same time, because China and India want to maintain control of their national situations, Wallace trains people on the ground, in the countries at hand, providing assistance especially with setting up labs, maintaining quality control, and conducting active disease surveillance. A major part of his mission was to get the CDC to recognize the regional importance of Southeast Asia. In response, the CDC has funneled in tens of millions for influenza research.

Although he had built a successful project with local collaborations painstakingly built across the region, Wallace was worried about the reliability of the stream of funding needed for projects that do not produce significant commercial outcomes. He noted that "Asian states are crisis-driven; they pour money into a crisis, and then turn attention elsewhere." Perhaps the same can be said of the United States. For instance during the SARS years, 2003–2004, infectious diseases were a major focus, but interest and funding have perhaps faded, except for research on dengue fever, which has returned as a major concern in recent years. Wallace's career illuminates one aspect of virtue in cosmopolitan science, not in terms of dedication to an abstract truth, but by being alert to invisible or emerging threats, to be vigilant when attention shifts elsewhere. He positions himself at the frontlines of what he calls "newly emerging infectious diseases" (NEIDs) that threaten the world.

So Wallace touches on the ambiguous links between crises and responses, and thus our inadvertent technological shaping of uncertainties. The attention span focused on particular medical crises defines whether the crisis is considered ephemeral or intermittent. The Nixon war on cancer also captures the mind-set of American leaders as they are driven from one crisis to the next, a minor pattern that shapes technological uncertainty while losing sight of the larger unknown lurking on the horizon.

These three American scientists, very different in sensibility and area of expertise, are involved in building new programs that can benefit society that did not exist before in Singapore. They partly left the United States during the time when there was controversy over stem cell research, neglect of research on infectious diseases, and cuts in federal funding for science. At the same time, they shared excitement about new biomedical data, funding resources, and opportunities to develop bioscience expertise in Asia. Of the three, Wallace is the best case that can be made for virtue in the interest of advancing the global battle against infectious diseases in a time of pandemics. He wants to revive a World War II American lead in tropical medicine, to return to Southeast Asia in the age of the jet plane. He wants to undertake research and spread skills for combating pathogens in Asian niches, that is, to lay the foundation in preparation for a world increasingly menaced by NEIDs. By the time of this write-up, the three scientists have stayed on, and they believe they are contributing to public and global interests even while drawing huge paychecks.

It is important as well to note that superstar scientists have a luxury perch in Singapore. During his decade as the leading spokesman of Biopolis, Edison Liu, a Chinese American, had a pied-à-terre in the Shangri-La Hotel, tucked in a leafy enclave off Orchard Road, the main shopping avenue. Beyond glamorous lodgings, global professionals are lured by Singapore's tax incentives—tax rates for foreign individuals are 20 percent and foreign corporations 17 percent.[16] Other well-remunerated expat scientists, besides Liu, can buy upscale condominiums and employ a couple of foreign maids to maintain their households and take care of dependents. Biopolis itself is well equipped with international restaurants and day care for young children of expat workers. Singapore is a strategic nexus of two streams of foreign workers—"world-class" scientists and "Asian" domestics[17]—very different in class and skills, but a great match for Western professionals seeking time out from doing with less and more uncertainty back home. In addition, the Changi airport, often rated the best in the world, allows for frequent quick getaways to sites in Asia.

"READING PROTEINS IS MORE IMPORTANT THAN STUDYING DNA"

Much younger foreign scientists also have something to gain by coming to Singapore. A UC Berkeley PhD, Dr. Sommers, with long tousled hair and a modest personality, was eager to tell me his story. After graduating from Berkeley, Sommers married a woman from Taiwan and decided to pursue his career in Asia. When I met him early in 2010, he was a postdoctoral fellow at SMART, a Singapore version of the MIT entrepreneurial model of assembling students, researchers, businessmen, and venture capitalists in an innovation center for the commercialization of ideas in order to create spin-offs. In the biosystem program, the goal was to develop new technologies that address critical medical and biological questions applicable to a variety of diseases, provide novel solutions for the health care industry, and generate a constant source of new technologies to the broader Singaporean research infrastructure. As a participant, Sommers said that many universities find it attractive to have links in Asia and that "Singapore was the ideal place to convert biotech innovations into products."

After all, Sommers received generous funding from NUS for his training. He was one of three PIs in a project to commercialize a computer tool for testing the blood by reading tiny units of proteins. He said, "Reading proteins is more important than studying DNA because proteins tell you what is going on right now. There are important pathways and signaling processes to capture." The project was a next-generation protein chip capable of detecting, from a drop of blood, hundreds of disease markers for cancers and heart and infectious diseases. For diagnostic and therapeutic purposes, he argued, protein studies work better than single nucleotide polymorphism (SNP) studies. "It is much like a music CD," he said. With his glass chip, "one can do a blood test and in twenty minutes . . . figure out a hundred diseases."

So when I asked him about genes and customized medicine, he answered definitively, "Personalized medicine is already here. It is called the blood test." The emphasis on ethnic-specific medicine, he believed, was "needed to sell the science idea to the public, and is therefore just a form of niche marketing." This was the view of a bioengineer, and many scientists (including those interviewed above) would disagree with him. The chip is an efficient way to conduct simultaneously a large number of tests, but the nature of the test itself is rapidly evolving, requiring the chip to be reprogrammed. The value of Sommers's chip is best realized by combining its readings with other kinds of lab data that would refine diagnoses and therapeutic practices. But Sommers's comments indicate the kinds of tensions bred by such work and the interdisciplinary conditions in which biotechnical and biomedical experiments are conducted.

The original patent for the protein chip was created at NUS by Professor Roger Kamm of MIT. To commercialize the chip, Sommers remarked, the project needed to get people with different skills, various talents, and access to different markets (money in the millions was needed). Early funding was provided by SMART and a mix of commercial grants. Thus, beyond developing prototypes, "the work involves developing a team and showing that you can sell the product, that is, add value, then attracting a biotech company to be interested in acquiring it." Because of risks for the company in the early stages of an experiment, researchers try to delay making licensing arrangements so that they can get more in the final division of profit. Also because SMART was a research entity in NUS (which owns the patent), his team had to negotiate the intellectual property rights, expecting "to get more than 50 percent or we won't be able to survive."

At the end of 2010, Sommers and his team founded a biotech startup in Singapore and in Germany, where he was born. He has won many awards since and has become the CEO of a biotech company with more than two dozen employees. Before all that glory, Sommers told me that spin-off companies provide a lot of benefits for Singapore by helping to raise and bring in capital and showcase successful companies. He noted that Berkeley too has lots of private institutions already doing this. U.S. universities like MIT are good in negotiating such deals with private companies; however, German universities do not do as well. Sommers was also grateful that there was a much bigger Singapore state role in supporting bio-entrepreneurs than in Europe. The Singapore government provides venture capital to support young companies like his for five to six years before they offer initial public offerings (IPOs). He noted that politicians in Singapore took great risks in funneling money toward biomedical research; the city-state seemed to have huge pension schemes and financial security. But despite all the advantages he reaped in Singapore, Sommers located the headquarters of his biotech company in Germany. So the question is whether foreign scientists can touch down to the degree that would serve people in Asia, or just their own careers.

Foreign Lab Workers

Sommers was exceptionally talented and entrepreneurial, with experiences at world-class universities and big companies. He confessed that negotiating licensing with global companies was more difficult than developing the protein chip. By contrast, the majority of mid-range scientists from Asia are not yet as entrepreneurial. Looking at PhDs recruited from Chinese universities, I hope to capture how different their experience is, one animated less by science entrepreneurialism than by lab drudgery.

Singapore has the highest concentration of global drug companies in Asia, and one of the attractions is that besides the excellent research infrastructure, they can recruit relatively inexpensive PhDs from China and India to work in their Singapore labs. Most of the science workers are from China, partly because there is greater uncertainty surrounding the certification of applicants from India, and who tend to be good in synthetic chemistry rather than in biomedical research. Chinese science students, in contrast, have a reputation for being strong in basic science, polymer chemistry, stem cells, and molecular biology. Wuhan University in particular and surrounding institutions in Central China are a hotbed of science talent that is tapped by Singapore.

The state sector has a well-coordinated program to recruit students and PhDs from China and India. The overall goal is to provide a so-called win-win situation for the researchers and for the government, and to inspire local talent. Besides institutes at Biopolis, NUS and Nanyang Technological University have developed fields in biosciences and state-of-the-art labs to attract foreign students from their giant neighbors. Singapore's multiethnic populations and cultures are promoted by headhunters as providing a compatible environment, education, and possibility of cultural rooting for ambitious science students from India, China, Taiwan, Sri Lanka, and across Asia.

A*STAR's global award program is aimed at the recruitment of young science talent worldwide. There are awards and scholarships for applicants at different points in their careers, from young students and undergraduates to postdoctoral fellows. Foreign undergraduate students are bonded and required to contribute to Singapore's economy upon graduation. Recently, I met a young woman from India who received an undergraduate science scholarship at NUS. After getting her degree, she spent three years in a science lab in fulfillment of the bond. She said that the Singapore scholarship program was mainly interested in recruiting "labor for labs," not in encouraging creativity in science. During her undergraduate years, she explored her interest in Indian dance and, after leaving the lab job, decided to enroll in a U.S. graduate program in performance. Was she going to leave Singapore then? To my surprise, she wanted to become a Singaporean citizen and contribute to sustaining authentic Indian dance traditions there. A decade after the Biopolis initiative, the program continues to recruit promising students to be future science workers. Sommers commented at a dinner with PRC postdocs that as "mid-level science workers they do important but underappreciated work at the labs."

The main focus is on recruiting foreign graduate students who get fellowships but without any strings attached. They are trained under supervisors who will help them develop their research skills and portfolios. Top researchers will be picked for research institutes at Biopolis, where each laboratory will receive yearly funding of up to US$500,000 (as of 2014). Other postdoctoral fellows can apply to work in corporate R&D labs, public administration, companies, and academia. A*STAR recognizes the critical importance of foreign scientific talents in forming the backbone of research and the need to groom the next generation of leaders. The goal is to induce outstanding foreign researchers (balancing the number of Chinese and Indian awardees to the official ethnic proportionality) to settle down in Singapore as citizens.

China produces hundreds of thousands of PhDs each year, and many of them seek to go overseas because of the greater demands of the job and higher salaries. The A*STAR fellowships have drawn over a hundred scholars and make it easy for them to become permanent residents and citizenships. When they have children, they may decide to become citizens, but it is not clear many will stay on. Some may view Singapore as a springboard to other places.

I had coffee with a couple of PRC graduate students, who seemed well adjusted and were considering whether to remain in Singapore. The young woman, Qing, was a second-year PhD candidate in internal medicine and auto-immunology. Her boyfriend, Han, was a first-year graduate student in pharmacology, both at NUS. Qing had a medical degree from Shanghai (an eight-year program) and had spent a year working in a government clinic there. As a student, she had wanted to go abroad, thinking, "If I get a U.S. degree, I can get a better job than at home." She took the GRE exam but was not accepted by any U.S. institution she applied to. Singapore responded earlier, gave a full scholarship, and provided a stipend of S$18,000 a year (in 2004). When she gets her NUS graduate degree, she will be able to work as a postdoctoral fellow in a state research lab.

Her friend Han had a bachelor of science degree from Wuhan University, majoring in biology. He had also applied to the United States and was accepted by Syracuse University. But despite three tries, he failed to get a visa; U.S. immigration wanted him to prove that he will return to China upon graduation. Han was unlucky because, following 9/11, U.S. embassies make it hard for students to obtain visas (leading to a significant decline in PRC students entering the United States, as compared to Europe). He also got into top universities in China such as Nanjing and Fudan universities, but at that time NUS recruiters visited Wuhan, and Han felt that Singapore was a chance to go overseas.

Both Qing and Han estimated that the yearly intake of foreign students at NUS was about two hundred students. Most of the students earning bachelor's degrees were from China and awarded scholarships but were subjected to a three-year bond for training in engineering, bioscience, and related fields. In Han's research lab, all five students were foreigners, while his supervising professor was an Indian expatriate. He reported that students from India also wanted to go to U.S. universities but, failing that, came to NUS as a second best choice. It appeared to Han and Qing that all grad students in the science fields were foreigners.

As foreign students, they found it easier to get along because of linguistic and cultural familiarity with Singaporeans and Malaysians. However, they regretted that it was hard to develop the kind of deep friendship they can have among mainland Chinese who have similar issues and interests. "PRC students want to live together and do things together." Upon getting their PhDs, PRC scholars have a few choices: find local jobs and get permanent residency in Singapore, return to China's booming economy, or go to the United States as postdoctoral fellows. The main factor influencing the decision of many to stay in Singapore, they said, was the kind of job they can get. By this, I think Qing and Han meant a high salary and career development prospects. "Otherwise," Qing said, "many are not keen to be in Singapore, complaining about it being so dull here, and the unchanging weather." At the same time, PRC postdoctoral workers realized that staying in Singapore for a few years of job experience provided leverage back home. Furthermore, with a Singaporean visa, they could achieve more flexibility as they could easily travel back and forth to China.

Meanwhile, the young scientists realize things back home were not that promising. In China, the yearly influx of students to colleges increased the size of the skilled labor market, thereby lowering salaries for graduates. Their friends back home could not even find jobs, especially in the big cities on the coast where they all wanted to live. Big salaries were possible for only those who had foreign experience. And foreign firms pay better too. Given the situation, English fluency provided an edge, and college students, Qing and Han claimed, spend one third of their time learning English. Qing mentions a friend in Shanghai, an overworked, poorly paid doctor who could not afford to buy a house. I left the couple with a sense that upon graduation and getting jobs, they would probably stay in Singapore after all. Qing said they were getting old and wanted to start a family soon. Although PRC fellows tend to be hardworking, reliable, and serious workers, the notion of simple virtue in science was very far from their pragmatic calculations. It was not a question of creativity versus drudgery, or academia or industry, but which global site offers the best conditions for highly trained workers to make a rewarding life. So Singapore's aggressive recruitment of PRC scientists may yet be sustainable, despite the tendency of many to consider the island a stepping-stone to a dream job in the United States.

Mainland Chinese workers in Biopolis labs seemed satisfied with their jobs and salaries, but they seldom get credit for the work they do. They seemed to appreciate the investment made by the Singaporean authorities in their talents, as well as the efforts to woo them to become citizens. They understood that there may be some grievance among locals because they are citizens and yet have less chance of getting scholarships to attend NUS, whereas all foreigners do.

A huge amount of money has been pumped into the biotech field, and there is general resentment against foreigners, especially if they are suspected of getting jobs that locals can perform. In order for the international talents program to really work, most of the foreign scientists are urged by the authorities to get permanent residency after one year. They also enjoy benefits such as getting housing loans, access to subsidized Housing Development Board apartments, and jobs for their spouses. They also receive exemption from the military service that all able-bodied Singaporean males must perform, as the army service obligation only applies to the second generation. One quarter of Singapore's over 5.5 million residents is made up of foreign expats and their families. General resentment of "foreign talent" in all areas—technology, science, education, but also drama and sports—continues, and the question is whether it will dissipate in time as Singapore becomes fully reengineered into a knowledge economy.

Local Science Talent

If Singaporeans have only basic degrees, they might
as well wash test tubes.
—Interview with a leader of A*STAR

When it came to a sustainable workforce, the long-term projection was that a cadre of citizen-scientists will eventually take over the Biopolis project. The main stumbling block among Singapore's hardworking students is that the majority were not interested in the drudgery of science work; most wanted to go into finance, which seems to promise a fast track to wealth. Singapore had very few PhDs in any field. Almost overnight, Singaporean youth had to be weaned away from basic degrees that only qualified them for washing test tubes, an official scolded, to aspire to serve their country as top-notch scientists. In partnership with the Ministry of Education, A*STAR tried to groom the young for a science career by intervening at different levels of education. There was a youth science program to cultivate interest in the sciences and to encourage talented youths to consider career paths in the fields of science and engineering. An award system sent promising high school students to work with scientists at Biopolis, to be followed up with short trips to labs in MIT and Johns Hopkins University in the United States. My two nieces were part of such recruitment exercises, including a two-week visit to Johns Hopkins, but after getting their college degrees at Yale, they ended up in banking jobs back in Singapore.

At the college level, talented Singaporeans are offered scholarships for overseas training in the life sciences at prestigious universities mainly in the United Kingdom but also the United States. Scholarship recipients are bonded for three to six years to return to work in Singapore. The figure named as an investment in each student sent overseas was half a million Singaporean dollars. The government had not stinted in its effort to nurture more young Singaporean scientists through A*STAR scholarships in order to build a significant pool of science expertise. After a decade of such an expensive strategy, the research community in Singapore (public and private) numbers over 5,400, with hundreds of clinical scientists. Half of the scientists are located at Biopolis. An employee at the JHS center commented, "If bioscience is viewed as the fourth pillar of the economy, the government cannot allow it to fail."

The suggestion is that fear of losing out, if nothing else, will harden the resolve to keep Biopolis as a high-profile public endeavor afloat. Nevertheless, it is a struggle to persuade Singapore's security-conscious students to consider an arduous career in biosciences. The *kiasu* anxiety that emerged in an earlier context of national survival in a hostile region has intensified in the globalized market environment. But the extreme form of striving that undergirds family relationships also breeds dread of failure or losing out to someone else. More recently, the personal fear of losing is tweaked by the need to compete with foreign students, state pressures to consider science an alternative to the big focus on business and finance in college, and the desire to not be channeled into jobs that are less promising for rapidly building up personal wealth. Many young Singaporeans aspire to be millionaires (in U.S. dollars) by the age of thirty.

In recent years, efforts to broaden the appeal of the life sciences and nurture more creativity have included the "global schoolhouse" initiative, whereby NUS and the National Technology University become partners of world-class universities.[18] The goal was to improve training in bioscience, computational systems, nanotechnology, and electrical engineering, areas that are driving the knowledge economy. More recently, in an effort to introduce American liberal arts education, Yale-NUS College was established in 2011, in the hope of fostering more creative and risk-taking younger generations. The schoolhouse project has attracted smart foreign students from surrounding countries as well as Singapore, thus compelling citizens to compete with the best students in Asia. Either way, whether homegrown or foreign, most science workers needed in the Biopolis ecosystem are routed or wish to be routed through institutions in the United States, their perceived home of cosmopolitan science.

There is another route for feeding Biopolis through institutionalized forms of not winning that work better on young men, partly because they are subjected more to cultural lessons on family and national duty. As mentioned above, all able-bodied Singaporean men spent two years in the army, and the armed forces are an important conduit for potential scientists who will work in the state sector. The Singaporean PIs (Yang, Tan, Ang, Ma) interviewed for this book have come up through a mix of academic and military science, pathways that provided ample opportunities to bind research activities to patriotic duty.

For all the efficacy of *kiasu* in shaping and driving a generation of scientist-citizens, Biopolis can seem to be one more system, like military service or bonded scholarships, that is more about creating a set of institutional constraints and mechanisms for managing human resources than it is about the cultivation of "civic virtue." As a kind of negative virtue, *kiasu* drives Singaporeans to try harder, lest they fail in multiple endeavors. At the same time, the compulsion for Singapore to win in this high-stakes science project is a nudge toward risk-taking. The Biopolis initiative is itself a gamble with uncertainties: a tiny city-state doing big science, the huge investments of tax dollars, brutal global competition in the pharmaceutical industry, and the capacity to fend off the next biothreat. Then there is uncertainty as well about how Singapore, drawn into the orbit of American science, thinks about its role.

Dr. Ang, a Singaporean native, was trained at Stanford University, receiving doctoral and medical college degrees. At that time the tendency was for Singaporeans to get higher education in the United Kingdom, not the United States, but the Singaporean government sent him to California. As an American-trained Singaporean, he provides an interesting contrast to the expatriate scientists, and he has an ambivalent perspective on U.S.-style medical training. When I met him, Ang repeated my question, "What made me decide to focus on cancers when I returned from the U.S."?

"Science is both a collaborative and a competitive field. Our choices are strategic in relation to a network. What area would make us eventually be on par with the top labs in the U.S. and Europe? We look at the number of disease conditions predominantly of concern to clinicians here, not so much in the U.S." Unsurprisingly, for government-supported scientists like Ang, a Singaporean vantage point made sense and was almost mandatory. Being positioned in the Biopolis system meant that the scientist competed against other biomedical centers in the world. He mentioned another Singaporean PI at the Genome Institute as wanting "to go head to head with MIT and Harvard" in stem cell biology. Singaporean researchers in that field also see themselves as competing with China, where "the stem cell science community is a powerhouse, using massive resources, et cetera. But they are not creative." Ang seemed to suggest that, compared with his expatriate colleagues, more was at stake for him as a Singaporean citizen-scientist at Biopolis.

I asked whether his American training fostered creativity in biomedical science. Here, Ang expressed ambivalence over the rapid push into American-style training for clinician-researchers. He noted it was not until the last decade that Singaporeans started embracing U.S. higher education. In the early years, some university administrators were against the expansion of Duke-NUS because it was too costly and trained clinician-researchers too quickly. One effect of Duke-NUS is to make medical students begin to specialize early in their training. By contrast, many still preferred the on-the-job training of the British medical system, so that doctors who were not pushed to do research have more time to give high-quality attention to their patients.

Ang commented, "The U.S. represents big science and transnational science, and having many collaborators. The question is, where do you want to sit? I want to be somewhere in the middle." He preferred science in small labs, comparing scientists to small businessmen, who can be thinkers, agile, and competitive. More broadly, talking about university education as a whole, he claimed that "we are now turning back to British higher education. Things are always changing here in regard to policy, and we are now in transit, wondering whether it is good to get a liberal arts degree. The problem is that it takes too much time. Also, the British approach to science is better; they sit, have tea, and mull over a question, and then come up with something creative." But despite his comments, the NUS system had pushed ahead to introduce American liberal arts education, including the Yale-NUS school.

Nevertheless, Ang made the point that people miss the British approach to medicine as being more humane and less corporate than the American one, liberal arts notwithstanding. As an oncologist, Ang investigated "an Asian-specific" type of stomach cancer that was a number two cancer killer worldwide. Because it was a "less well characterized cancer, we look at Option D: genetic polymorphism, infection, and diet." But like many of his Singaporean-born colleagues, Ang was involved in other projects prompted by his two-year stint in the army. His medical training delayed his military service, so that when he enrolled in the military, he was assigned not to the field but to defense research.

During training in Thailand, he found healthy eighteen-year-olds were dying from infection by a soil bacterium. Ang identified the cause in a CDC biowarfare Category B agent, a bacterium that causes the Melioidosis infection, an acute and chronic form manifested in joint pain, skin infections, and pneumonia. Rice-paddy farmers were most at risk of getting this disease. "It's a cool bug, a very complex microorganism endemic in Thailand, and much of Southeast Asia." Ang collaborated with Thai scientists at Mahidol University (a Wellcome Trust outpost) in Kon Kean. He pointed out, "This was all before Bush's focus on biowarfare . . . although Americans in Vietnam suffered from the disease and it was called the Vietnamese time bomb or Vietnamese TB." This story is Ang's way of underlining the point that, for him, bioscience was mainly about service to the nation and people in the region. He seems to imply that a kind of human-centered approach that focuses on small groups and creativity is best. Instead of speeding ahead in the U.S. version of entrepreneurial science, he felt that British-style medicine would provide time to "cogitate" about what kind of medical culture Singapore should pursue. In a separate interview, Ang's compatriot Ma said, "It's a question of how the nation wants to position itself in the future." Where does virtue in bioscience sit in Biopolis with regards to balancing the health needs of people in the region with upholding "scientifically responsible ethics"?

This chapter has argued that as biosciences become relentlessly entrepreneurial and global, scientific virtue is not only context-dependent but varies with the different categories of science workers. Biopolis, as a milieu of cosmopolitan science, illuminates how (at least) two different formations of virtue are at play when foreign and locally born scientists adjudicate among values of doing ethically responsible science. Expatriate scientists and their Singaporean colleagues tend to raise different kinds of questions of what is at stake for "the common good," itself an unstable category. Whereas mobile researchers seem more driven by technological gains that can bolster their careers and benefit the biological life of humanity at large, Asian citizen-experts must wrestle with questions about how the interplay of biotechnologies and biopolitics shapes the life of biological, social, and citizen subjects in the region.

Globally, European and American science leaders have expressed concerns about how unfettered commercialized science threatens disinterested inquiries and how "scientifically responsible ethics" needs to be charismatically exercised and upheld by PIs. As we saw in the above examples, foreign PIs in Biopolis's "science nirvana" practice exemplary science by engaging in regulated experiments, designing tools, and building programs that can contribute to humankind. In this way, scientific responsibility to various populations with different national, genetic, and biological vulnerabilities intersects with and is conditioned through the life plans and career pathways of individual superstar scientists calculating these two timelines in tandem. There is a propulsive, individualistic dimension to the expatriate scientist's trajectory that is about making a name in a chosen field, including by training physician-researchers in new areas.

Among the expat scientists whom I interviewed, only two also frame their scientific contribution to their home nations. Wallace's leadership of the Duke-NUS infectious diseases program hopes to build back up the American presence in tropical medicine in Asia. Sommers, the young German bioengineer, wants his invention to benefit and open up German bioscience, and he leverages Singapore's research as a means to achieve this goal. But given the established and sizable science communities of the United States and Germany, Wallace and Sommers are free-floating scientists who can decide where to do their work as a form of disinterested inquiry. By comparison, for many PRC and Indian lab technicians, science is mainly a job opportunity to access an overseas market. Landing in Singapore's well-funded labs, most still aspire to gain entry into North American or European research institutions.

For Singaporean-born scientists, ethical decisions are much more burdened with the moral obligations of being citizens and being "Asians" in an era of emergence. Science that matters is about negotiating between the old and new versions of the Asian nation-state. Because their work is tethered to the island's bioscience aspirations, local scientists are expected to exercise their civic duty by inspiring, recruiting, and training younger Asian scientists in order to sustain Biopolis's responsibilities. As core members of a regional community of science experts, they must shape and defend ethical norms of scientifically responsible research that is and will be conducted in developing countries. Their moral leadership includes the practical and ethical weighing of health problems to be dealt with and experimental projects to pursue, that is, shaping notions of the "common good" in this particular region. For instance, should Biopolis continue in its relentless march toward the goal of designing blockbuster drugs or take a slower, more considered approach to deadly diseases that menace people in Asia? Personal virtues in a regionally oriented bioscience would require modulating dreams of fortune and fame (potentially more easily realized by moving into the private sector, or by going abroad) and shouldering the scientific responsibilities of emerging nations. Their commitment to this larger enterprise means that the fear of failure is itself a virtue in shaping questions of how to live as scientists in and of Asia.

SIX

PERTURBING LIFE

"The genome is more plastic than we thought," said Dr. Ma. "We can reprogram the entire genome and push it to an earlier stage." We spoke in April 2010, when I visited the Singapore Genome Institute to speak to one of its luminaries. Ma, a Singaporean scientist who trained at the University of Edinburgh, referred to a new technique that induces stem cells from differentiated adult cells, thereby bypassing the use of human embryonic stem cells. In 2006, scientists started experimenting with different principles of acquiring novel cell states by inducing mature (somatic) cells to develop into different cell types. As biologists who spend long hours fiddling with proteins, Ma and his colleagues perhaps also think biologically about medical applications of stem cell technologies as a way to improve society.

With a new capacity to "modulate cell fate," in Dr. Ma's words, scientists can also induce a broader sociopolitical pluripotency that has huge implications for bioscience in Asia. As Ma put it, the question is, "Can we program these applications for human benefit?" I understood that he meant for humanity at large, but implicitly also the Singaporean and Asian competition in this daunting science field.

In classical Greek mythology, Zeus punished Prometheus by chaining him to a rock for his theft of fire. Each day, an eagle was sent to pick at Prometheus's liver, which, by regrowing every night, allowed the eagle to continue the torture in perpetuity. Stem cell research has the ambition to be a Promethean science. By using a patient's own cells to regenerate diseased tissues, thus circumventing immunological problems, researchers hold up a vision of the human body as a self-curing machine. This Promethean quest to unlock the secrets of cell replication and versatility has different kinds of biomedical application. It raises questions about immunology itself that extend to questions about the negotiation of what is human and what is machine in the body, and ultimately about whether immunology will become a kind of politics as well.

As Niklas Luhmann remarks, "We speak of risk only when and insofar as consequences result from decisions."[1] For instance, the decision to modulate immunology in order to avoid risks of rejecting novel therapies ripples across different fields. In an essay on the immune system as a figure for a xenophobic politics, Donna Haraway argues that immunology practices reproduce ideological notions of selfhood by modeling medicine that is premised on an "us-versus-them" conflict. She explains that the immune system operates as a biological metaphor for the determination of what belongs and what does not, ultimately defaulting to a fantasy of a self-same body disentangled from others.[2] As I will suggest below, the uncertain implications of immunotechnologies are bigger than a reinforcement of our individualistic tendencies and may have geopolitical effects besides.

Immunology is a branch of experimental medicine, whereas immunity is a phenomenon, and the immune system refers to the physiological processes that regulate an organism's health. Stem cell technology views the body as an autopoietic system, but it comes to articulate other self-organizing systems, such as the artificial organ, and beyond the lab, by extension analogically to the research milieu and society at large. Thus, by developing induced pluripotent stem (iPS) cell technology and associated therapeutic interventions, cell scientists at Biopolis are perturbing life in multiple registers, accelerating the race—both in the sense of competition but also of biologized segmentations in the human species—of science and a rather geopolitical rivalry in scientific versatility.

Modeling Autopoiesis

In biology, autopoiesis refers to the self-reproduction and self-maintenance by a system from within, or by using its own cells. As a biotechnology, autopoiesis refers to the technical self-organization of the system due to decisions made by experts to sustain its capacity to reproduce and also to regenerate and autoregulate.[3] Biotechnological autopoietic systems deal with incoming risks by modulating their negative influences on systemic self-reproduction and self-regeneration. At Biopolis, I observed two related experimental cell techniques that modulate risks to immune health. First, iPS cell technology aims to generate a supply of undifferentiated stem cells not from using material from human embryos, but from reprogramming adult (somatic) cells to revert to an immature, versatile state. In the iPS cell technique, researchers introduce chemicals to the cellular system and successfully induce the reversion of adult cells into stem cells that can grow tissues for different types of organs. In Singapore, Ma commented that cell scientists can now "turn on or shut down genes at will" as well as halt the pluripotency of diseased cells. The critical point is that iPS cells generated from a patient's own body can potentially be used in therapies without triggering a negative immune reaction.

At the moment of this writing, the therapies are not standardized or generalized to this degree. Induced pluripotent stem cell science and its vision as articulated above are still in the experimental stage or trial. The therapeutic endeavors and hopes will carry into the next generation, by which time other futures and values may be articulated.

But for now, the experiments are under way, though therapies are not always standardized across labs in the world. Researchers focus on the use of iPS cells to form artificial organs in vitro outside the patient's body that can recapitulate and mimic organ development, structure, and function as they occur in vivo within the patient's body. For instance, scientists use iPS cells to grow diseased organoids so that they can study tumors and test drugs in the lab before novel therapies can be safely transferred to the actual body of the patient. By modulating cell fate, cell biologists can also modulate the immune system of the patient. The patient's body itself is thereby operationalized as a self-curing equipment that uses its own iPS cells as a novel therapeutic technique. By culturing tumor cells, cell biologists grow beach-ball-size organs in vitro to model diseases in vivo. The two techniques—inducing cell pluripotency and culturing artificial organs—enhance the versatility of the human body, recasting it as both a self-curing machine for improving health and a guinea pig for the lab (as a model for itself).

Pluripotent cells and immunotechnologies can generate a cascade of effects and events within and across systems. Modulating stem cells and immunology engenders cascading effects in the understanding of the human body, such as blurring the difference between the vulnerable body and self-repairing machine, between living tissue and a stem cell factory. It also sparks a string of effects in consideration of the ethical conundrums and objections to all of the above. While immunological research reframes the human body as an autopoietic system, in addition, the research milieu itself comes to be operated as an autopoietic system, functioning in a self-correcting, self-generating way in the context of global ethical debates and cosmopolitan science.

At least in Singapore, there seems to be the imagination or the extension of the experimental iPS experimental system to the level of an experimental culture of bioscience. In this high-stakes field, expertise in cell versatility and disease modeling is vital for sustaining Biopolis in the brutal competition for national prestige. It has become an important grounds for attaining recognition for Asian scientists and perhaps, in the future, gaining the Nobel Prize. With the transnational connectedness of multiple systems, from the scale of the autopoietic body to the international race for scientific prestige, will perturbations of life cascade into the realm of immunopolitics in the future?

I begin by describing how the trajectory of cosmopolitan cell biology has been oriented by a recent string of events linked to innovations by Asian scientists, including those at the Genome Institute of Singapore. In the West, ethical debates over the use of embryonic stem cells have contributed to a perceived slowness in stem cell research that cell scientists in Asia hope to turn to their advantage in the race to promote their national profile. Next, I trace how ethical issues have shifted from reproductive labor to the potential contribution of stem cell technologies for curing childhood diseases. I then turn to Biopolis, where stem cell research is focused on modeling diseases, projecting the body as a self-curing machine, and banking ethnic-differentiated stem cells as strategies to make immunology the future of post-genomics. The final section analyzes why researchers in Singapore consider stem cell research an "Asian" arena, tracing the stakes of competition with scientists of Asian ancestry in other sites. The resonance of "Asia," as in genetic variation, immunology, and cell expertise, particularizes research ambitions to develop autologous therapies for fast-growing aging populations in the region. In my concluding remarks, I mention some broader effects of immunological science, the uncertainties surrounding the idea of the body as a self-curing machine, and the time and monetary commitments that can be sustained especially in Singapore.

Transitions in Cell Biology

> We are in an era in which we generate
> stem cells at will and with ease.
>
> —DR. MA

In the late 1990s, researchers dreamed that human embryonic stem (hES) cells might be used to repair damaged cells, thus creating a field known as regenerative medicine. Early research on the cells, which were to be procured from gestating human embryos, triggered a huge ethical minefield, especially in the United States.[4] In the midst of extended ethical objections, the U.S. government limited funding except for research on a few "approved" cell lines. Some of these cell lines were provided by Dr. Arif Bongsu, director of the in-vitro fertilization program at the National University of Singapore (NUS). In 1984, Bongsu was among the first researchers to grow hES cells on human feeder cells rather than animal feeder cells. Indeed, Bongsu's coup was the critical impetus that launched the Biopolis program in cell biology.

In 2006, another Asian scientist, Dr. Shinya Yamanaka, helped to effect another shift beyond the early intense public focus on human embryonic cells. Working at a lab in Kyoto, Yamanaka developed a technique that induces iPS cells from adult cells derived from lab mice. iPS cell technology opened up a new field of exciting science and its projection of a leap in stem cell medicine, promising a supply of stem cells for therapeutic purposes without the need for human embryos. There is worldwide competition to produce effective iPS cells for novel therapies for Parkinson's, Alzheimer's, Lou Gehrig's, diabetes, and other diseases associated with aging. In classical physiology, or twentieth-century biology, stem cells, which are undifferentiated cells, were found in embryos and related fetal substances only. With iPS cell technology, scientists can now make stem cells out of mature cells that can have remarkable potential to develop into many different kinds of cells.

Yamanaka, who seems bound for a Nobel Prize (he has won the Kyoto Prize), "single-handedly shifted attention to Asia," Edison Liu noted. In Singapore, the Yamanaka coup has had an electrifying effect on the small group of stem cell scientists who view the timing and place of the invention as a clear indication that iPS cell research is an "Asian" field. This is especially true as powerful American scientific institutions, in their estimation, missed this opportunity while tangled in bioethical and philosophical debates about procuring the cells. In some ways, stem cell technology seems a unified answer to two major problems in Asia: (1) the pressures to improve livestock via cloning in order to adequately feed huge populations, and (2) developing stem cell therapies to treat their high rates of degenerative diseases. With such practical urgency and nationalist winds behind the science, there is also a relative lack of controversy surrounding the conduct of cell biology (see below).

Yamanaka's pioneering of the technique in Japan has thus been interpreted in Singapore as a testament not only to the importance of Asian researchers globally, but also to the special character of stem cell research as an "Asian" field. This is the perception despite the fact that Yamanaka has since moved part-time to the University of California San Francisco (UCSF) Medical School and has set up a stem cell program in Kyoto University. Researchers pointed to cell scientists of "Asian" ancestry (i.e., researchers from Japan, Korea, Taiwan, the People's Republic of China [PRC], as well as Asian Americans) who can be found in major labs in the West. Thus, cell scientists in Asia see iPS cell research as a uniquely "Asian" arena, no matter where Asian scientists are located in the world.

Ma's lab at Biopolis was competing to improve on Yamanaka's method by shifting from animal to human experimentation. He framed his experiment this way: "What makes a stem cell a stem cell? How does one make a non-stem cell a stem cell?" First, Ma and colleagues use proteins to perturb cell regulatory networks in order to "jump-start ES [embryonic stem] cell-specific expression program in somatic cells."[5] With iPS cells derived from lab mice, Ma had "created chimerical mouselike animals that showed the production of organs" that were later sacrificed.[6] However, the technique is intricate and uncertain, and experiments in animal studies indicate that the viruses used to introduce the reprogramming factors into adult cells can cause cancers. The race is on to induce stemness in adult cells in mice so that this process works in humans as well, without viral outcomes. This is extremely daunting, time-consuming work, and researchers continue to search for or develop the right model organism for applications in humans.

Ma also wanted to speed up Yamanaka's technique by reprogramming adult cells for an optimal induction of iPS cells, as well as for them to perform as well as hES cells in laboratory conditions. If laboratories succeed in inducing adult human cells to "adopt a pluripotent fate," then researchers are closer to their ambition to take it to the next level, that is, to use reprogrammed stem cells to repair damaged tissues in humans and to grow organs outside of bodies for research and therapeutic purposes. His lab seems poised on the cusp of something momentous, yet great uncertainties remain in both ethical and therapeutic areas. Meanwhile, contrary to some assumptions, hES cell research has not thereby become irrelevant.

Temporalities

The ethical maelstrom surrounding human embryonic research in the West has fueled the media reception of iPS cell technology, which sidesteps concerns over the ethical status of embryonic cells, as the end of hES cell research. Although there is a promising new method to generate stem cells, human embryonic research continues to flourish, sustained in part by iPS cell technology that has borrowed its methods and materials. After all, work on hES cells has been foundational to the invention of iPS cell technology.

In an interview in California, Yamanaka said that the iPS cell technique circumvents "the ethical controversy—we have to destroy embryos to isolate embryonic stem cells." While this suggests that embryonic stem cells are becoming obsolete in stem cell biology, a few lines later, he said, "Embryonic stem cells are still important for the development of iPS cell research." Indeed, iPS cell research had progressed quickly because of findings from hES cell research, including methods to create various types of cells. Yamanaka points out, "That's why iPS cell research has evolved so rapidly." Diplomatically, he suggested that delays in stem cell innovations in the West are "partly because the use of the cells has caused an ethical controversy around the world."[7]

There is therefore a material/technical continuity between iPS and hES cell research. At the same time, scientists at Biopolis claim that the intertwining temporalities of stem cell research and the ethical uproar in the West worked to their advantage. Not only did iPS cell technology emerge in East Asia, but the hiatus created by bioethical and religious concerns in the United States did not happen in Singapore. Ethical criticisms of American embryonic stem cell research in the 1990s, Asian scientists note, provided them with a critical lead. Under the George W. Bush administration, religious objections to the use of human blastocysts for research led to restrictions on the availability of human stem cell lines and to less federal funding for public universities and institutions working in this field. Because of lawsuits, the California Institute for Regenerative Medicine did not start researching until 2006. Obviously, private institutions such as Harvard and Stanford that do not depend on federal funding were never hindered in their pursuit of research. Today, because of the media celebration of iPS cell technology that does not involve using human embryos, ethical debates surrounding the field for the most part faded away or shifted in tenor.

Shaking their heads over the use of bioscience as an "ideological football" in the West, Asian biomedical experts claim that they have a head-start over international stem cell centers. At the height of the controversies over cell research in the West, a few scientists such as Dr. Alan Colman, the famous Scottish cell biologist who cloned Dolly the sheep, left for Singapore. Scientists there are convinced that they can maintain their lead in time and in the development of innovative cell research. In Singapore, coordinated state support for science, a social ethos of inducing consent via consultation of community leaders together with a lack of public protests create a stable environment for cell research. A semijoke going around expatriate circles claimed that the best thing about Biopolis, besides the sparkling labs and generous funding, was that research will not be disrupted by animal rights activists "who care more about animals than people" (more below).

As discussed earlier (in chapter 2), a national Bioethics Advisory Committee (BAC) was appointed in 2002 in order to forestall any possible ethical objections from a pretty complacent population. Biopolis quickly established itself as a center for stem cell research, unhampered by the delays in the United States. The government recognized the immense potentials to be generated from stem cell research and quickly adopted existing British legislation for embryonic cell research, which included a ban on the use of human embryos beyond fourteen days of age as well as a prohibition on human cloning.[8] Because of its robust embryonic cells program, Biopolis researchers have sped up training in iPS cell work for the time being.

The claim is that because Singapore has dealt effectively with the ethical issue of hES cell research, Biopolis's project is not distracted by political bickering. Dr. Ma explained that researchers need a combination of the two technologies (iPS cells and hES cells) in order to learn how to reprogram cells to repair damaged tissues in the human body. ES cells are used as research material for the iPS study of cell differentiation and as control (comparison) materials; they are important for testing the quality of iPS cells. One can only presume that a sizable portion of the hES cells is mobilized, via the use of the consent form, from embryos and fetal tissues donated to local hospitals and fertility clinics for research and potentially therapeutic values.

Ethical Uncertainties

> Stem cells are not a magic cure for the old,
> but most promising for babies and kids; to
> restrict stem cell research is unethical.
> —DR. ROBERT WILLIAMSON,
> at the Singapore-Australia Symposium
> on Stem Cells and Bioimaging, Biopolis,
> May 24, 2010

Stem cell research has raised many ethical uncertainties, from the harvesting of reproductive labor to a misleading magic cure for old age. Feminists have focused on the ethical concerns regarding the role of tissue donors. Sarah Franklin observed that, due to the increasing proximity of embryo donations and stem cell research in the United Kingdom, there is a "double reproductive value" involved in the flows and exchanges, mediated by technologies of consent and regulation.[9] More recently, Melinda Cooper and Catherine Waldby argue that cell biology critically entails a kind of invisible "biolabor" performed by female donors of reproductive tissues. The maternal-fetus nexus is a necessary reproductive precondition that permits lab scientists "to reanimate and revalorize what might otherwise be wasted life, channeling it into uncanny potentials."[10] Elsewhere, I have discussed the commercial and social pressures in Singapore for mothers to donate "hospital waste" for research and to bank their infants' cord blood as a practice of techno-savvy parents.[11] Indeed, "adult stem cells" are derived not from adult humans, as commonly believed, but from fetuses and cord blood. Because only ES cells are able to recapitulate all gene activity that engenders pluripotency, they are still superior to iPS cells. But the clinical use of hES cells is particularly hazardous because of safety risks like the potential formation of cancers.

Thus, for cell scientists, the moral charge of using voluntarily donated reproductive materials weighs heavily as they struggle to come up with safer and better techniques that may eventually benefit humankind. The need to clarify and shape public perception continues, and the ethical conundrums are by no means settled. Furthermore, media-generated "ethical evangelicalism" (as scientists disparagingly call it) misled the public by misrepresenting stem cells as a potential cure for some celebrities, such as the late actor Christopher Reeve and President Ronald Reagan, who had various degenerative diseases. According to ES scientists, the form given to ethical controversy will further delay the pace of innovation in stem cell research. Ethical debates in the United States and other places such as Australia also do not understand that stem cell research holds the most promise for childhood, not old age, diseases.

Stem cell scientists around the world gain a boost from Yamanaka's technique, and the media attention helped stimulate more flows of money into a field that requires a long timeline to get robust cell-based products to the market. Since the work on recombinant DNA in 1995 in London, scientists have been doing stem cell research in sites around the world. Led by Colman, other British stem cell scientists have emigrated to Singapore and the antipodes for warmer climes when it comes to research facilities, more generous funding, and less bureaucratized and audit management via internal review boards than might be found in the United Kingdom. Not surprisingly, disgruntled cell scientists find Singapore a freer venue in which to vent their frustrations and discuss how to correct media views about what seems to the public to be a Faustian bargain.

"Human genetics is *so* last era," proclaimed Dr. Robert Williamson, a British expatriate scientist based in Australia, at a Biopolis stem cell conference in 2010. He explained that "the two new areas of science are the brain and the first few weeks in human development. The two questions rely on stem cell research. Thus ST has a commanding role; we need to distribute knowledge of this science." Because "stem cells are not a magic cure for the old, but most promising for babies and kids, to restrict SC research is unethical."[12] In regenerative medicine, then, cell scientists are trying to find a unified theory of everything that goes wrong with aging bodies from the moment of birth. They boil all problems down to the lack of stem cell responses in the many diseases—Alzheimer's, Huntington's, strokes, heart diseases, broken bones—that a fresh infusion of human stem cells can cure, or so it is hoped. The larger claim is that aside from ethical unease, it would be unethical to deny stem cell research that holds out promising therapies for babies.

In its usual pragmatic fashion to take the shortest, least controversial route, Biopolis has elected to focus on pursuing human iPS cell technology as providing potential treatments for genetic diseases associated with aging people, a huge group already identified as the intended beneficiaries of stem cell technology. The main approach is to use stem cells to model diseases outside the patient's body. By turning iPS cells into nerve cells, clinicians can study genetic diseases in the heart, liver, brain, or other organs. Human iPS cells generated from patients displaying genetic diseases are used as a surrogate model to understand some of the early molecular correlates of that disease (i.e., a disease phenotype or in vitro condition). For instance, Colman has been working on modeling human diseases such as insulin-dependent diabetes and cardiovascular diseases. His regenerative medicine company, ES Cell International (ESI), is forging a stem cell research network with Japan, China, and Australia. There is an emergent regionalization in stem cell research as immunology becomes the frontier of cosmopolitan science.

The Body as Machine and Lab Surrogate

> Immunology is going to be big; it is the
> future of post-genomics.
> —EDISON LIU, 2010

In response to a question about "the most important stem cell breakthrough of all time," Yamanaka's answer was that "the ability to generate an unlimited supply of patient-specific stem cells will revolutionize the future of medicine."[13] An

American professor visiting Biopolis elaborated: "We are at a point when the cell biologist wants to make every cell he can get hold of into a stem cell. We are at a time when we develop pluripotent cells easily in the lab."[14] Indeed, by modulating the "stemness or pluripotency" of cells in human beings, scientists have taken the first step toward achieving new techniques that promise that, one day, the human body can make its own stem cell medicine without the hazards of rejection from its own immune system. Thus the goal of iPS cell technology is not (just) to use stem cells to regenerate damaged cells but to induce alternative forms of human cell pluripotency in order to model human diseases through the patients' own cells.

Technical and ethical uncertainties dog the production of a diseased alter-body to serve as a lab surrogate for the patient. To avoid immune-related issues, this living model of the sick patient requires customized stem cells to test whether the patient's own cells can self-reorganize and be self-sustaining. Such autologous therapies test drugs on the surrogate model before being introduced into the proper body of the patient. Thus, iPS cells are intended for inducing from the patient's own cells a surrogate living, albeit diseased, body alongside the patient's one. iPS cell technology is thus the basis for a whole new sphere of research on immune-resistant therapies (including organ transplantation) that enrolls the patient's body in its own cure. In the future, doctors will nurture our diseased twin bodies as guinea pigs, to be tested for medicines so that treatments are safe for us.

Such exquisite customization of the body's own immune system and genetic makeup means that clinicians will benefit from the access to a spectrum of stem cell diversity, which in Singapore means they will be ethnic-specific. Novel stem cell therapies are "patient and population-specific medicine" because the degenerative diseases that they target have a basis in the personal or particular mutations. The Singapore strategy is to accumulate an "Asian" source of stem cell as a great investment that may one day yield potential values from close research on differentiated stem cell lines. Demystifying the process, Ma said, "Stem cells are little delivery factories that secrete proteins for patients." The Biopolis reasoning is as follows. Because the generation of iPS cells from each patient in accordance with the best methods would be expensive, lengthy, and difficult, there is a need for the establishment of an iPS cell bank, to collect unique iPS cell lines that are homozygous for major human populations.[15] At Biopolis, the iPS cell bank will focus on cell lines from "Asian" people because most of the beneficiaries of yet-to-be-developed therapies will also be largely of Asian ancestry. The gathering in of heterogeneous stem cells will greatly support genetic diagnostics and the modeling of major "Asian" genetic diseases. But all that lies in the future, and meanwhile scientists need to also find secure funding for their painstaking research.

Meanwhile, the yield of research leading to the treatment of major aging diseases is extremely slow. So far in Singapore, autologous transplants have included cultivated human conjunctival stem cell explants and limbal stem cell explants. The testing of stem cell treatments for eye diseases is significant because it is a prevalent genetic problem among East Asians and is considered a threat to intelligence and, thus, a potential weakest link in Singapore's focus on intellectual knowledge. Also, work on eyes and retina regeneration has been reconceptualized by stem cell research around the study of brain diseases and neural stem cells.

Once again, although such degenerative diseases are common to all humanity, there is the perceived Singapore edge in focusing on their manifestations among peoples with Asian DNA. Central to the novel therapeutics are the recognition of stem cell heterogeneity (cf. genetic variation, cancer heterogeneity) and the need to use the patient's own stem cells. As a tool to modulate cell fate, iPS cell technology also, by implication, modulates the fate of the patient seemingly destined to die from a range of diseases associated with aging, and also the fate of large Asian collectives susceptible to major inheritable diseases. Because the application of stem cell therapy is patient-specific, such customized medicine can be viewed as also "population" or ethnic specific. "We are, however," Ma added, "at the early stage of the technology, and it is one fraught with difficulty and competition," especially from other cell scientists who also happen to be "Asian."

Intra-Asian Rivalry

The life sciences today are without borders and are thus an arena of fierce competition, increasingly geopolitical in nature. One measure of the competitiveness of the biomedical world is the number of big corporations making the rounds in Biopolis and in Asia at large. I attended a few conferences held at Biopolis where global firms (specializing in pharma, equipment, international law) sponsored lectures and distributed colorful brochures, thumb drives, and other souvenirs. Noting "the deep biological information" possessed by Asian researchers, a representative of Sigma-Aldrich, a medical equipment company, claimed that the "key stem cell scientists are in Asia."[16] Whatever the value of this corporate-driven claim, cell scientists in Singapore definitely agree that Asian cell scientists are their main competitors.

Stem cell research, many claim, is both an Asian enterprise *and* an arena of intra-Asian scientific competition. There is a disciplinary perception that Asian scientists are highly represented in this field. Yamanaka's landmark work was not a fluke, but reflected a strong trend in stem cell research in East Asia, partly based on animal research and breeding. In 2004, there was a South Korean attempt to repeat "therapeutic cloning" (somatic cell nuclear transfer, or SCNT), the method of using an adult cell to replace the nucleus of an egg, within humans. Professor Hwang Woo Suk, a veterinarian, was accused of making false claims about creating the world's first SCNT cell lines for human cloning. The international scandal that ensued, including about his unethical acquisition of human eggs, eclipsed any idea of Korean sophistication in stem cell science. Dr. Hwang has since capitalized on his skills at cloning family pets, thus creating a global pet-cloning industry.[17] In Japan, the ethical debate around hES cell research has been muted as well. The "opaque regulation" surrounding therapeutic cloning may have created conditions for Yamanaka's development of iPS cell technology.[18]

In China, scientists have also excelled in animal cloning, driven in part by a nationalist project to improve livestock to feed a gigantic nation.[19] President Xi Jinping started out in pig farming and had visited Iowa to learn breeding techniques from American farmers. Meanwhile, iPS cell research is racing ahead on different fronts, including research on Chinese stem cells and their potential uses in therapies and transplantation. Chinese expertise in animal cloning has contributed to the creation of hybrid lab animals. In Singapore, I heard a PRC scientist report on his experiment with ovarian stem cells in mice and subsequently cloning chimerical mice that reproduced themselves for at least two generations. Taiwan also has a robust program in cell research.[20]

Above, I mentioned the significant contributions of Arif Bongsu and his group in Singapore that first derived hES cells from five-day-old human blastocysts (obtained with informed consent) from local fertility clinics.[21] According to Colman, Bongsu's work paved the way for Dr. Jamie Thomson at the University of Wisconsin to produce long-lived hES cell cultures, a development that triggered excitement in the late 1990s over the role of stem cells in regenerative medicine.[22] Given this legacy, local researchers consider Singapore the most appropriate site for iPS cell work for the treatment of degenerative diseases for patients with Asian DNA and mutations.

There are, however, multiple avenues for reprogramming somatic cells that are being pursued by scientists in Asia. Ma sees himself as improving on Yamanaka by coming up with a more efficient method (using proteins and genes) for converting fibroblasts into pluripotent cells, whereas at the Scripps Institute in California, PRC-born scientist Sheng Ding examines how molecules control cell fate at the metabolic level. At the same time, scientists are exploring cells from different animals, especially fish and salamanders, both of which have very high generative capacities. At Stanford University, scientists (including one of South Asian ancestry) have discovered a method to reprogram one type of adult cell without having to convert it into a stem cell (thus avoiding the danger of forming tumors), for the purpose of repairing organs. Clearly, cell technology is moving along multiple pathways, and the competition is as intense and difficult as any in the history of modern biology.

Meanwhile, Singapore cell scientists believe they have other advantages vis-à-vis their counterparts elsewhere. Here again, Biopolis researchers enumerate their pluripotency in this field. In the United States, many scientists have to apply for federal funding, which is a real "turn-off," compared to the kind of block funding structure that sustains Biopolis. There was no mention of Britain, which Sarah Franklin argues has the most "favorable economic" conditions for stem cell research in the world.[23] Positioned outside the transatlantic sphere, Singapore, with its English fluency, is perhaps more cosmopolitan and globally oriented than the science cultures of Korea and Japan. China, I was told, is beset by the lack of scientific regulation of work, especially by China-trained scientists. Ma expressed his view that Singapore was still ahead of China in stem cell research and that most of his postdoctoral students are from the PRC. Mainland Chinese scientists with world-class training in the West are fine collaborators, and also competitors. India does not yet have a stem cell research presence. Thus, when Biopolis cell scientists say "Asian," they almost always mean "East Asian," adding Singapore to the mix.

In the field of "Asian" cell biology, there was a semiserious joke that "Asian" scientists were endowed with exceptional "chopsticks" skills to manipulate stem cells, to the extent that Charis Thompson dubs the South Korean milieu "Hwang's chopstick pastorale."[24] There is an echo here of another discourse common decades earlier when Asia was drawn into global manufacturing. The recruitment drives of American and Japanese factories in Malaysia, I discovered, proliferated narratives about the "nimble fingers" (attributed to biological difference and cultural training) of young Asian women that make them especially dexterous and patient for high-tech assembly work.[25] In the current bioscience endeavor, culturally specific masculine talents are at stake here, and the narrative of chopsticks (akin to calligraphic brushes in Confucian cultures) indexes the fierce nationalist rivalry between stem cell centers in Asia, but also between them and "Asian" scientists who are located in the West. Researchers at Biopolis point to the many cell scientists with Asian origins who work in labs in Boston, San Francisco, and Ann Arbor. This perception of a global distribution of "Asian" stem cell scientists frames not only the identity of scientists, but also a specific set of skills, as well as the transnational field of cell expertise, as ethnically "Asian."

Scientists in Singapore (and other sites in Asia) believe that the direct beneficiaries of yet-to-be-developed stem cell therapies will also be huge populations largely of Asian ancestry. Therefore, Singapore is building a stem cell bank that carries differentiated Asian stem cell lines for research. Ma told me that his stem cell work "explains how individuals are different. We look at a certain DNA sequence: What are the risks or prospects of getting certain diseases?" His achievement is the development of patient-specific iPS cells as treatment, especially to advance disease modeling or to push iPS cells into an earlier stage in order to find what went wrong. Thus, making adult cells adopt "a pluripotent fate" may also be a strategy for persons sharing the same kinds of disease-causing mutations. At Biopolis, the stem cell scientists view themselves in competition with their counterparts everywhere, with a focus on the mainly PRC-born scientists in the United States, especially those at the Scripps Research Institute in La Jolla, California.[26] In other words, stem cell research is fast emerging as a global field dominated by Asian scientists in Asia and in the diaspora.

"Asian" from the Cell Up

What are the ethical implications of "an era in which we generate stem cells at will and with ease"? Researchers are themselves stymied by such questions, and they try to skirt the larger questions raised by the sophisticated tools for creating, identifying, and modulating cell pluripotency. They point to strict ethical guidelines in Singapore-based research (no use of human embryos older than fourteen days old; no cloning whatsoever) and the potential of developing cell therapies for major diseases menacing human beings, Asians in particular.

Some circumvent the ethical implications by drawing attention to the experimental nature of stem cell knowledge. For instance, Ma cautioned that stem cell technology "is not mature. It may not be perfect. We are still working at the upstream process, refining the technology." There are, however, new scenarios we have never encountered before, including "the capacity to immortalize cells simply by scraping some from the skin." Whereas in the

United States a concern has been about stolen cells that later earn labs huge profits, a situation considered almost impossible in Singapore,[27] the issue that Ma raises is the potential capacity of researchers to immortalize their own cells. He then fell back on the dilemma between "knowledge is a black box" and the ethical self-surveillance of the researcher. "Scientists do this kind of work because of passion. We have ethical boundaries and understand when to stop. Scientists are human too, and they have their own ethical conscience." I interpret Ma's comments about work and passion as an upholding of science virtue, as well as civic virtue (scientists knowing when to stop and not to do harm); the ethics of being a human being and a scientist give him the authority to decide how science experiments can benefit society. But he did not seem disturbed by the idea of the body as a stem cell factory. Indeed, an "Asian" stem cell bank is needed because stem cell diversity may be linked to micro-perturbations that signal differences between ethnic groups. But again, ethnic-specified stem cells as the source of autologous therapies still lie in the future, and the scientist counseled patience.

In keeping with doubts about stem cell therapies (researchers avoid the term "regenerative medicine," which they consider misleading), the time frame is of necessity projected as lengthy and uncertain. Biopolis scientists guess that there is an average twenty-five-year window for a possible stem cell cure and that the cell program will need US$25 billion during this time frame. They believe that Singapore is where "long-term state commitment" is more likely than in other countries. But in late 2010, the Singapore Ministry of Trade cut back state support for the Genome Institute and other research institutes, requiring scientists to find outside finding and industrial partners. So even at Biopolis, cell scientists now have to work hard to obtain money to fund their Promethean question.

There is thus a perilous balance between hope and potentiality, time and money, in stem cell research. Perhaps there, in this long-distant race, the narrative of Asianness—"Asian" innovations, the head-start in in vitro fertilization, the potential benefits for customizing Asian-specific genetic diseases, the glory of Asian scientific prowess on the global stage—weighs against the huge costs to be borne by taxpayers in this high-stakes project. In this era of genetic diseases, I was told that a good place to start would be blood diseases that afflict "Asian" people disproportionately. iPS cell technology is after all about fiddling with proteins and cells in an ever-enlarging, circulating diaspora.

Above, I have traced the ways in which stem cell research emerges as "Asian." This includes the different histories and expertise in animal breeding and in vitro fertilization techniques that make Yamanaka's iPS discovery something that could be attributed to "Asia" rather than to a personal genius of the Venter type. Another "Asian" dimension in the field is its lack of an ethical hiatus compared to the West. In addition, the number and distribution of Asian stem cell scientists in Asia and in the West are significant. Finally, the development of iPS cell therapies, if ever realized, would create intensely personal medicine because the body's own cellular materials become the substance of its own repair. Meanwhile, Singapore needs to assemble an "Asia-specific" stem cell bank in order to carry out painstaking research on different cell lines. Such research materials, focusing on Asian cell materials, promise to one day spawn autologous therapies for major genetic diseases that affect vast aging populations in the region. Thus, through the interweaving of technological, molecular, corporeal, epidemiological, and epistemological dimensions, the "Asian" approach and "Asianness" of the knowledge is a continuous thread that connects all the other chapters in this book.

The Promethean Promise

I have argued that the exploration and development of pluripotency itself is not only a property of cells but also a property of the institutional, social, and experimental milieu being created.

By stealing pluripotency as a "natural" condition of embryonic life and making it a lab technique to artificially induce pluripotency from adult cells, Yamanaka has set off waves that rock the life sciences. I transposed pluripotency from biology to the larger Biopolis experiment itself as a key dynamic that animates science practices that seek to unleash values from singularities into multiplicities and to shape autoimmunities in multiple registers. But as I have tried to show, competition among global sites of cell research has been modulated by temporalities and uncertainties that generate in turn a cascade of effects that have human, temporal, budgetary, and political dimensions.

First, "pluripotency" as a science project of Biopolis is shaped by Singapore's/Asia's aversion to "ideology" and ethical debates that hindered the progress of research in public-funded labs in the West. Stem cell technology and immunology science modulate how researchers make decisions, and in Asian research hubs, the focus is on noncontroversial experiments focused on "pragmatic" projects, such as livestock improvements and treating aging-associated diseases. The Singapore trajectory prioritizes the development of therapies for human degenerative diseases, and bypassing the ethically dangerous research in hES cells may lead to new therapies for childhood diseases. Where there are ethical controversies, the Singapore strategy is to do what may be most unobjectionable, but cell therapies for aging diseases are more likely to succeed in the short run. The area of immunological treatments for aging bodies is where Singaporean cell scientists can make a mark, while responding to health and market needs in the region.

Second, can humans ever be induced to adopt a pluripotent fate? A unified paradigm of disease suggests that all illnesses are due to faulty cells, and these can be repaired by perturbing them. The body itself becomes a stem cell factory as well as a machine with many moving, replaceable parts, any of which can be made from any other. Furthermore, this self-curing body-machine can re-configure the immune system, thus modulating autoimmunity to accept micro-adjustments. Pluripotency thus projects a new concept of life as a perpetual re-setting of the clock, the body as a self-repairing machine promising the fountain of youth by redirecting biologically induced "beginnings" into longevity.

Third, stem cell therapies for inducing a self-curing body require the mod-eling of diseases. This involves the generation of a lab surrogate body that can be tested in place of the original body of the patient. The double-body technology—one merely diseased, the other a disease model—is an eerie conflation of the body-machine. What are the ethical implications of using ethnicized stem cells to model diseases for research on "Asian" bodies? Perturbations in stem cell research thus ripple across multiple temporal and material dimensions, from reverse engineering of cells to an embryonic state, to the externalization of the patient's tissue as a diseased form in the lab, to the Promethean promise of a self-renewing body, to the resurgence of Asian sci-ence on the basis of ethnic immunology. But then what of the eternal torment that self-curing makes possible for Prometheus, where the body's capacity for perpetual self-repair is precisely a form of divine punishment, and organ re-generation can be a fate worse than death?

Will Singapore's, and its rival Asian, experimentations with immunology give rise to a form of immunopolitics,[28] whereby pluripotent technologies shape views of nations as self-sustaining systems? If indeed the Singapore bio-science project's transnational scope can be sustained in the foreseeable future, one wonders whether an immunitary conception of biopolitics would shape a kind of *communitas* that, in multiple ways, could attempt to self-immunize against contamination from elsewhere? What other meaning of Asianness would emerge?

PART III

KNOWN UNKNOWNS

SEVEN

A SINGLE WAVE

> When we found out that all of humanity was derived
> from a migration out of Africa, it reversed centuries of
> Eurocentrism. That all Asians probably came through
> Southeast Asia and migrated northwards, once again
> brings us closer together, conceptually, as one people.
> —EDISON LIU

In 2009, the publication of a report on genetic diversity in Asia sparked claims about "a single migratory wave" of modern humans out of Africa. Edison Liu, a leader of the study, exulted: "In Asia, we are all related." The rise of Asian genetics, he asserted, has brought "us closer together, conceptually, as one people."[1] The plotting of a single entry point into Asia is framed as a (re)conception that weaves diverse genetic flows into a vast Asian unity. By setting into motion affects that circulate among scientific communities in different Asian nations, the authors of the population genetics project hoped to shape as well the intangible realities out of which an "Asian" unity in science can be mustered.

This first ever Asia-wide effort to collect DNA from indigenous groups can be traced to the Human Genome Project. Arguably the founder of population genetics, Stanford professor Luca Cavalli-Sforza helped launch a worldwide effort to collect the genes of contemporary native populations in order to answer questions about human migrations over the last hundred thousand years.[2] In Asia, Liu and colleagues, including former students of Cavalli-Sforza, launched the Pan-Asian SNP Consortium under the supervision of the Human Genome Organization (HUGO), a London-based international scientific research project to identify and map all the genes of the human genome. A team of Asian scientists came together at HUGO Asia to investigate single-nucleotide polymorphisms (SNPs), or variations at individual bases that make up the genetic code, in multiple populations in Asia. A decade later, the consortium published a working hypothesis in *Science* magazine claiming that a single wave of ancient migrations peopled the Asian continent.

At stake in the pan-Asian migration claims was a recasting of the anthropology of premodern Asia in its unified multiplicities beyond a "Eurocentric" perspective. As we shall see, new work on population genetics, conducted mostly by Asian researchers, challenges a commonly accepted "Eurocentric" theory of a two-pronged process of early human migration into Asia. Liu and his Asian colleagues argue that the new population genetics findings indicate instead a "single-wave" arrival of early modern humans on the continent. As an alternate version to the Eurocentric two-pronged model, Asian geneticists hope that this story of a unified ancient history will inspire a past-present-future sense of Asian unity in or through modern science. In their problematization, a region famous for its political fractiousness seems to require a framing of the continent as a special genomic population and collective disease afflictions. In this reconfiguration, they invoke shifting notions of "continent," "genetic origins," "population," and "disease," for tackling what are coming more and more to constitute something like a trans-Asian health and its incumbent challenges, now and in the future.

The genomic revolution is in part a revolution in identity, not only of individuals and groups, but also of continents and emerging regions in the world. The algorithms that sort and standardize desequenced DNA and genomic data are not just a diagnostic construct of genetic variations among diverse populations in the world. Genetic maps can be read as well as reconfiguring the polytemporal horizon (from premodern past to uncertain future) of genetic accumulation, analysis, and self-knowledge. This chapter argues that the trans-Asian SNP network as both a research infrastructure and trans-Asian cultural narrative is not just about promoting regional chauvinism, but about becoming prepared to intervene in a region rife with disease threats.

Preparation technologies, Andrew Lakoff has argued, include the technique of scenario making to generate an "affect of urgency" among the authorities regarding impending biothreats. Scenario-based exercises allow experts to model, construct, and then enact potential futures, with the effect on possible future-oriented political action in the present. The inter-Asian SNP project can be viewed as a scenario-making exercise that varies from Lakoff's model in a few instances. First, whereas Lakoff looks at "disease as a national security threat,"[3] there is an increasing need to generate a sense of disease as a *regional* biosecurity threat, especially in tropical Asia where infectious diseases are prevalent. For Singapore, there is the need, in a politically fragmented and unevenly developed region, to stir up affect and action across multiple national sites. Second, the timeline of preparation technologies is not present-future, but past-present as well. Especially in Asia, deep historical roots have to be tapped in order to underpin a contemporary imagination of sharing the present-future. This backward-looking detour is a necessary first step in a scenario-building exercise that hopes to influence cross-border action in the present-future. Third, therefore, *affective* forecasting, rather than scenario making, stirs less a sense of urgency than of genetic pride in belonging to a distinctive continent and civilization.

The Pan-Asian SNP umbrella has to be considered in both its infrastructural and narrative dimensions, the latter being more powerfully potent, perhaps. As a transborder health technology, Liu explained, analyzing ancient DNA in "migrant histories converging in the present will help us to account for mutations in Asian environments." As a novel figuration of biological space-time, the inter-Asia DNA database is a form of affective forecasting that uses genetic evidence of collective pasts in order to generate a renewed sense of pan-Asian solidarity. The discovery of primordial affinities locked in genomic codes can set powerful affects into play for sustaining the growth of regional science, among other things. Thus, whereas medical genetics taps into fears of vulnerability and hope, population genetics, in reimagining an ancient DNA unity, stirs affects that, circulating in a region notorious for its Balkan-like divisions, fosters a novel kind of collective reimagination in preparation for the challenges ahead in combating nonparticipation and threats to public health.

The outline of this chapter is as follows. First, I situate Asian science stories based on genetic findings as a form of affective forecasting that hopes to encourage cross-border scientific collaborations. Next, I explore why scientists invoke "pan-Asianism," a term that has a particular twentieth-century resonance in modern Asian imaginaries of a redemptive civilization. Third, scientists explain how the genetic evidence of ancient migrations allows them to recast "the anthropology of Asia." Genetic findings allow scientists to tell an

"Asian story" that stirs genetic pride in their convergent past as many peoples on a singular continent. Finally, I analyze how Singapore enhances its role as a mediator between Asian megastates by making affective projections of science pan-Asianism. In a final statement, I share my unease with this reconceptualization of Asians as brand-new ancients.

Scientists as Storytellers

Scientists are covert storytellers, and what more exciting story is there than genomic variability in the teeming Asian continent? In *Primate Visions,* Donna Haraway observes, "Biology is the fiction appropriate to objects called organisms; biology fashions the facts 'discovered' from organic beings. . . . Both the scientist and the organism are actors in a story-telling practice." By making "visible" the metaphor or math in "the life sciences as a story-telling craft," Haraway shows how, even as its objects are consequential in shaping the narratives about them, biology "appropriates life forms to fashion facts and make truth claims." She deconstructs primate studies as a mode of scientific writing that deploys "third world" animals as surrogates for cultural human "nature," touching on topics from sexuality to reproduction, from monogamy to patriarchy.[4] In a related vein, Jenny Reardon views the practice of embedding racialized scientific objects in genomic science as a technique that gives scientists the power to define race and democracy and to embed whatever goals of governance, including racist ones.[5] Both observers write from the vantage points of big science, where sexist and racist ideas are indeed present in shaping the knowledge of Western scientists. More recently, Jonathan Marks notes that "genetics, like biological anthropology, legitimizes a set of origin narratives—our micro- and macro-revolutionary ancestries, respectively—and consequently often has had tense relations with folk ideologies that seem to draw legitimacy from the science."[6] Knowledge formation, therefore, weaves an entangled matrix in which science and culture, materiality and metaphor, technicality and affect inexorably implicate and propel one another.

From the emergent biomedical frontiers, science stories are resolutely future-oriented and thus less about using science to legitimize old cultural ideas and more about using science to generate new imaginaries of a collective future. Storytelling in Asian science thus aims less to recover primordial origins in Asia as a means of bolstering an insular regionalism in the present, and more to create the narrative conditions through which to imagine an Asian future. When population geneticists in Asia talk about the past (genetic evidence gleaned from ancient human migrations), they are concerned about how continuities in race and ethnicity may have prefigured a wished-for contemporary unity, one that is always in the making and yet to come. Because population genetics is less concerned with disease and phenotype and just looks at how genetically diverse ancient peoples in Asia were, the distribution of genetic variations can be read as evidence of convergent cultural origins in the continent. The origin story proposed by population geneticists in Asia, therefore, is not entirely an affirmation of folk narratives but the creation of a whole new discourse of a single departure out of Africa that can generate an affect of solidarity for science cooperation in a region deeply divided by populations, histories, and cultures.

Therefore, whereas Euro-American geneticists tend to tell stories about life in general, such as DNA is a "river out of Eden"[7] or "the blueprint of life," researchers in Asia do not view genetics merely as anonymous digital flows but as a dramatic story of particular peoples whose historical migrations have shaped biological and cultural diversity entangled in a special region. An alternative to the view of DNA blindly self-replicating beyond its human hosts (e.g., Richard Dawkins's "selfish gene"[8]) is a big Asian story that finds, in its shifting patterns of DNA, clues to past-present and present-future phenomena at the regional level. Ironically, perhaps inspired by Dawkins's concept of a meme (a kind of cultural DNA), Asian scientists, who read the ribbons of SNPs that trace the ancient flows of humans into Asia, develop a story about a distinctive context of human intermingling and thus continental cultural distinctiveness. By interpreting population genetics as evidence of a single wave out of Africa, Asian scientists find compelling ancient lessons for a contemporary vision of an emerging scientific order or nomos, one that has the potential to advance life sciences in the region.

Life Sciences and Pan-Asianism?

The Biopolis ecosystem has engendered a cascade of scales for intervening into health problems looming on the horizon. At the micro-scale, fungibility is about making DNA, disease susceptibility, or a risk profile and standardized customized medical interventions somewhat interchangeable. At the regional scale, fungibility is about aligning cosmopolitan science with regional modernity, a goal that defies conventional planning and may require flowery language about Asian history, uniqueness, and future possibilities. The scale surpasses the nation-state as an imagined community based in reconstructions of primordial history in addition to a nationalist structure of feeling to gestures at an Asian future presaged by a deep anthropological unity traced through a relentlessly regional conception of human evolution.

In one of his more candid moments when he stepped back from his science spokesman role, Liu, the humanist scientist, rejected the mantra of biotechnology as a driver for the economy. After all, it has not turned out to be a major job or revenue creator for Singapore compared with other segments of the economy (financial services and casinos). He remarked that "biotechnology today is like a flower, a metaphor of cultural advances, the achievement of a level of living that includes clean air, green spaces." Here he identifies a potentiality, a piece of cultural and symbolic capital, a flourishing environment imagined and predicated both on modern Asian history and contemporary Asian science. As the would-be leader in this larger goal of preparing for the uncertain future, the Singaporean bioeconomy faces huge uncertainties stemming from the uneven development of bioscience and political rivalry that beset Asia.

Why does Biopolis see itself as a potential leader in this seemingly impossible endeavor? Much of it has to do precisely with the position of Singapore as a cosmopolitan entrepôt in Asia without some of the more problematic international relations concerns of larger regional powers. For instance, Japan, the first modernized nation in Asia, has not played a leadership role in biomedical science for the region. Deep political wounds inflicted by the Japanese invasion of nearby countries during World War II in the name of "Asian co-prosperity" keep the nation from playing a leadership role in the region. After the war, modern science in postwar Japan has been focused on the domestic context. Liu notes, for instance, that the pharmaceutical industry is based on an "old boys' network" coddled by the Japanese government. He says that Japanese national protectionism of the biomedical sciences means that "there has been no development of a global rigor in competitiveness, compared to, say, the international prominence of the Japanese automobile companies." Nationalist concerns have also dominated the development of biomedical expertise in South Korea (see chapter 6) and China, where despite the global success of a giant DNA sequencing company (see chapter 9), the biomedical scene there is "chaotic; anything can happen because it lacks scientific regulation." India has not yet built the necessary scientific infrastructure, and Indian firms tend to be focused on the manufacturing and factory production of pharmaceuticals, not research. In larger Asian countries where science is a nationalist project, the tendency is to collaborate with scientists in the United States and Europe and to view other Asian nations as competitors. Meanwhile other Southeast

Asian nations do not have serious science expertise but are also focused on solving pressing domestic problems of public health.

Furthermore, when it comes to genomic science, most countries view DNA as a new bioresource and symbol of national vitality, and they have become reluctant to share genetic information. From China to India, South Korea, Vietnam, Indonesia, and Malaysia, DNA is highly regulated and not meant for export. Biosovereignty has been expressed in different ways. In Japan, genomic materials ensure the nation's capacity to build its own pharmaceutical industry (China is catching up), and Indonesia has been perhaps unfairly viewed as a country that ties disease samples to commerce.[9]

Nevertheless, researchers in Biopolis have been adept at fostering overseas research collaborations even when the DNA samples never leave the home country, as is the case with China and India. Singapore does not have such limits on the outflow of human tissues but does implement international guidelines for the safe transportation and use of data. It has also been strict in observing international standards for research. Singaporean multicultural success in forging research relationships with other Asian countries seems to position the tiny island as a fairly neutral mediator between the megastates of China and India, and between technologically advanced centers and less-developed Asian countries. Michael M. J. Fischer as well has observed that key researchers at Biopolis also contribute to the new life science economy by engaging in "science diplomacy."[10] Indeed, Biopolis seems to be the place to promote the sharing of local clinical, human, and epidemiological samples and data from different Asian nations, to build the kind of science sharing and goodwill that is necessary for a potential coming together in the future, perhaps to confront looming biothreats.

It is striking that the inter-Asian genomics network invokes "pan-Asian" regionalism, a modern concept that emerged from inter-Asian political ideological struggles against Western domination in the first half of the twentieth century. Prasenjit Duara argues that early twentieth-century political and cultural Asian movements promoted a discourse of worldly redemption, incorporating older ideals of spiritual civilization as a source of moral authority. Unfortunately, in 1940, Japanese imperialism appropriated pan-Asianism, using "Asia for Asians" as a euphemism to justify its brutal invasion and occupation of Asian neighbors during World War II.[11] But the notion of pan-Asianism (traceable to China's founder Sun Yat-sen's "Greater Asianism," or *Da Yaxiyazhuyi*) continues to offer civilization as a supplement to nationalism. Since the 1980s, the economic rise of many Asian countries has created different forms of regionalism (e.g., Asia-Pacific Economic Cooperation or APEC). Building on these networks, HUGO Asia hoped to promote scientific collaborations for shaping an inter-Asian research region. Science pan-Asianism therefore is not just about configuring a "metageography,"[12] but about deploying novel discourses of history, race, and culture in order to conjure up a new imaginary of inter-Asian values. The Asian SNP network created opportunities for scientists to tell gripping stories about braided ancient routes and roots, hoping with the aid of the news media to play up their conjoined political, cultural, and ethical dimensions. Against what they consider the Eurocentric rhetoric, the main goal of scientific pan-Asianism is to rally different science communities on the continent to cooperate for redemptive ethical ends.

To Liu and other Asian scientists, the great tragedy of a biologically rich but fragmented region stems from its politically poor showing of actual trans-Asian, crossethnic collaborations around biosecurity. Especially in the aftermath of severe acute respiratory syndrome (SARS), and the spread of infectious diseases such as avian flu and dengue fever, international experts had a different vision of the future in which Asian countries ought to come together or at least coordinate knowledge and action around health threats that slip across borders.

What if algorithms tracking affinities among diverse peoples in a vast region can be deployed to recast Asian peoples, in all their diversity, as having a common ancestry? ("Unity in diversity" is a governing ethos in Indonesia.) Can truth claims about Asian DNA engender affects of commonality and urgency as a first step toward scientific problem solving at a regional level? In other words, Biopolis, as a champion of cosmopolitan science, seems a reasonable site for switching on expressions of pan-Asianism that can rally an urgent sense of health vulnerabilities and the need for redemptive action.

Recasting the Anthropology of Asia

Scientific entrepreneurialism includes the capacity to shape good stories out of pretty challenging materials. HUGO Asia scientists telling stories are expressing a mix of scientific opportunism, *ressentiment*, paleo-fantasy, and, especially for Biopolis promoters, pragmatic insights as to how to get a region-wide genomic project rolling. For scientists involved in the SNP network, the initial challenge was to pry the Asian genomic picture from the one created by the Human Genome Project, which ignored the fact that peoples in Asia form more than half of the world's population. But immediately the main challenge was how can a vast region, fragmented by diverse populations, languages, religions, politics, and economic circumstances, be brought together to build a regional Asian DNA database? How can stories about a unified DNA past-present incite a scientific friendship for building a research infrastructure around a novel sense of "Asia"?

In 1998, HUGO cataloged the common variants in European, African, and East Asian genomes.[13] As Liu remarked to me, "There is of course no such thing as a single human genome." Almost immediately, leading Asian scientists sought to improve upon the original Hapmap project. Yoshiyuki Sakaki, the president of HUGO Asia, said that it was "a long-term dream of HUGO Asian Pacific" to found a trans-Asia network called the Pan-Asian SNP Consortium to study genetic variations on the continent.[14] Because of Liu's leading role in founding the Asian SNP network, he became the president of HUGO (2007–2012), moving its headquarters from London to Singapore. By 2011, the HUGO Pan-Asian SNP Consortium had conducted the largest survey to date of human genetic diversity among Asians. This Asian genotyping database is an online research resource that contributes to the study of human diversity. The resulting information on the ancient flows of DNA hints at the historical convergences and divergences that came to be interpreted as genetic evidence of a common "Asian" ancestry. The storytelling aspects of pan-Asian network-ing animate collaboration among researchers in different sites by tapping into a shared primeval past.

In population genetics, researchers tend to deploy different concepts of "race" that define the biological and nonbiological features associated with a group. American scientists prefer the term "ancestry" or "ethnicity" to "race" for the identification of biological and environmental conditions that influence a person's genetic variation.[15] Similarly, in Southeast Asian genomics, the nonbiological concept of ethnicity is deployed to define groups through their membership in a linguistic family, by geographical proximity, by a known history of admixture, and by isolated settlements, all factors indicating "similar ancestry estimates."[16] There is thus a mixture of scientific and cultural classifications that together seeks to capture both genetic and epigenetic elements in groups defined through ethnicity or ancestry.

I was repeatedly told that race and ethnicity are research tools for pinpointing the gene–environment interactions that produce genetic variations in susceptibility to certain diseases and response to treatment. At a New York conference, Liu remarks that "race and genetics are uncoupled, while race and environment are highly coupled. . . . We should segregate populations for study based on genetics rather than race." However, in order to obtain statistical significance, "the oversampling of minority [ethnic] populations is necessary to enable identification of effects." The question is how to isolate genetic clusters in specific environments that are linked to diseases? Race becomes a mechanism to pinpoint genetic pool or environmental exposure, "to identify the social, economic, psychological, and biological variations in the human community."[17]

In my interview with Dr. Svensen, Liu's former colleague and an American population geneticist now at the University of California at San Francisco, explained that "ethnicity is a research tool rather than a claim about racial difference; it permits research controls on the basis of ethnicity." Indeed, the racist charges that arose from the Human Genome Project drove quite a few population geneticists to Singapore, where there is a pragmatic view that ethnicity is a convenient mechanism for a first-line mapping of the distribution of genetic variation and disease predisposition, understood as the outcome of a long and complicated evolutionary, migratory, and demographic history. Svensen himself, as a scholar at Stanford University, was affected by a signal conflict within the campus department of anthropology, which split up because of an unresolved disagreement between paleoanthropology and research in evolutionary biology, on the one hand, and sociocultural anthropology, on the other. There are also disciplinary fights between paleontologists and paleobotanists versus bioarchaeologists and primatologists as their struggle to build up or tear down empires continues. In Asia, scientists, as products perhaps of a less contentious, high modern Western science, tend to dismiss much of the furor as ideological battles that obstruct critical scientific work.

But then in Asia, distinctive physical traits and histories have their own kind of emotion-laden significance that cannot be banished in the sterile world of statistically generated metrics. Liu explained how the SNP network was a "relatively benign first step" for grasping human diversity in the region. Population genetics is after all merely "a simple experiment, a first pass at how diverse Asians are." Mapping migrant histories converging in the present, in other words, is a kind of anthropology that traces the origins and identities of peoples in Asia through their DNA lines. It is clear that the DNA-driven anthropology is not merely about cultural diversity, but is basically an infrastructure to shape a new scientific geography that is also "Asian."

The SNP project sampled seventy-three populations scattered throughout eleven countries, with about two thousand samples covering a wide spectrum of linguistic and ethnic diversity. Populations, identified by modern ethnic labels, included aboriginal Negritos, Karens, Aeta, and Uyghurs as well as Northern and Southern Han, Javanese, Koreans, and "upper-caste Indians."[18] Besides the samples, the scientists themselves came from many sites. The SNP network began with the membership of eleven countries, including China, India, Indonesia, Japan, South Korea, Malaysia, Nepal, the Philippines, Singapore, Taiwan, and Thailand, then expanded to seventeen member nations from the Asia-Pacific region. Thus, besides the goal of mapping genetic diversity across Asia, scientists from different nations were brought together to build an open database. The project was not meant to be comprehensive across the region, but by reworking the theory of genetic phylogeny, Asian scientists hoped to discover afresh the prehistory of the continent,[19] and, in this way, to genetically recast the anthropology of Asia by shifting from aboriginal stories of particularity to Asian stories of ancestral unity.

An Asian Story

The collection of multiple DNA samples in a single Asian genetic database created the moral basis for a revised anthropology of Asia. Paleoanthropologists and linguists have yet to weigh in to consider how the genetic findings may fit with existing fossil, linguistic, and archaeological evidence. But in the narratives of SNP geneticists, what mattered was not just past evidence, but also what the genetic code flowing through the continent seemed to foretell, a cultural story of the present-future.

Consortium leaders lost no time in asserting that the findings reveal hidden truths about the particularities of populations within a singular stream that came to constitute Asia as a natural and political whole. The single-wave hypothesis was established by repudiating earlier claims about divergent human flows into Asia. Conventional views based on mitochondrial DNA held that Asia was populated by at least two waves of migration from the north to the south. One model traced migration from mainland China through Taiwan into Southeast Asia and the Pacific islands. The other route came from Central Asia into the rest of the continent.

By contrast, the Asian SNP project reported a single "migratory swoop" some 60,000 to 75,000 years ago out of Africa into Asia.[20] Moving from south to north, this single migration hugged the coast of India and traveled through Southeast Asia before moving north into the Asian continent. A science reporter at Biopolis, Vikrant Kumar, articulates a novel claim for public consumption. This single-wave theory, he argues, identifies Southeast Asia "as the major geographic source of East Asian and North Asian populations," thus indicating a shared common ancestry.[21]

As may be expected, elite scientists from different Asian countries had different reactions to the hypothesis of the peopling of the continent. At the press coverage of a HUGO meeting in France, in the summer of 2010, elation and dismay were expressed in equal measure. Geneticists from their respective nations expressed dissenting views. An Indian scientist exclaimed, "All Asians arise from Indians"; a Korean counterpart remarked, "This is all bunk"; and a mainland Chinese scientist said, "We don't accept the SNP claim about Asian origins in Southeast Asia." Liu, who recounted such comments, noted that the value of the SNP "out of Africa" findings was that they turned old racial concepts, as well as the racial chauvinisms of the big powers, on their head. "Diversity is good because we now know about genetics." Taking a dig at what he considered the "sheer racism" of the elitist cultural view of East Asian big powers, he noted with satisfaction, "What was strong is now weak."

Indian geneticists, by contrast, enthusiastically embraced the evidence of India being the single entry point of modern humans into Asia. Indian media, including the *Times of India*, picked up the news by using phrases such as "ancestors of Chinese came from India" and "the Chinese evolved from Indians." Liu modified such claims by saying, "It is probably more accurate to say that Dravidians [in Southern India] and Chinese had common ancestors, than to say that Chinese ancestors originated in India."[22] A more circumspect view was expressed by an Indian scientist, Professor Samir Brahmachari, a member of the SNP project. He emphasized scientific cooperation over one-upmanship. "We have breached political and ideological boundaries to show that the people of Asia are linked by a unifying genetic thread. It was exciting to work together with scientists of such high caliber with contrary views and of diverse background—kudos to Prof. Ed Liu."[23] Indeed, for the scientists, the point was not to stir up a genetic war between Asia's two big giants, but to build an intra-Asian scientific community by sharing DNA data, resources, and expertise.

"A Coming of Age"

The focus on building a scientific infrastructure in Asia was not missed in global publications as well. A *Science* report notes that, with its ninety researchers in forty institutions in eleven countries, the project marked "a coming-of-age for the continent's genomic sciences." It cited Liu's emphasis that the project "was conceived by Asians in Asia and executed, funded, and completed by an Asian consortium."[24] Such emotion-laden claims seemed to suggest that the most fundamental achievement of the project, for its conceivers, was perhaps less in its findings than in the arrangements of transnational, intraregional scientific links that it precipitates: the formation of a scientific infrastructure that mimics and doubles the "Asia" projected by the genetic mapping.

Representatives of less scientifically developed nations spoke of hopes for overcoming political mistrust and accessing state-of-the-art technology that was beyond their reach. A researcher from Malaysia noted, "The PASNP consortium has shown that a scientific collaboration by scientists from diverse culture, race, religion and economic strata is possible." Another Malaysia-based scientist emphasized working together to answer questions "about the fundamental genetics of 60% of the world's population." The Filipino representative lauded the "collaboration between scientists from developing and developed countries," while the Thai member praised "scientific cooperation . . . beyond the limitation of politics and economic constraints of member countries." An Indonesian molecular biologist notes that, "for [Indonesia], a huge archipelago with more than 500 ethnic populations, such data is of public health importance and has had an immediate impact in the study of disease distribution." The Indian member emphasized the building of "bridges among population geneticists of various Asian countries."[25]

While poorer countries expected the transfer of skills and technologies from the SNP consortium, scientists from more advanced Asian nations highlighted the sharing of diverse DNA information. Comments by mainland Chinese researchers emphasized the spreading of "computational approaches" and studies of "high density markers." Japanese and South Korean members focused on the need to extend "the number and samples and ethnic groups" and "sharing data of each ethnic group." A South Korean scientist summed up the expectation that the pan-Asian initiative was a timely and "important resource for the nest level of genome research in Asia and beyond. The consortium has acquired key experiences of sampling, deposition, coordination, bioinformatic processing, and interpreting a tremendous amount of data."[26]

In other words, countries on the economic spectrum saw different kinds of value from participating in the SNP network. For less-developed Asian countries, the transfer of knowledge and skills from more advanced sites promises to overcome differences in resources across countries. For the richer countries, gathering data from across the continent would produce a pan-Asian genetic repository that will come to represent the shared ownership of the new science.

Singapore as a Switching Station

Narratives of common Asian origins help position Singapore as a switching station, operating switches that regulate genes by turning some on and others off. Whatever the scientific implications of population genetics data accumulated by the SNP project, the most immediate effect was to generate an affect

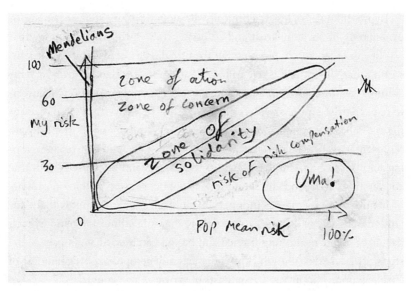

FIG. 7.1 A napkin doodle attributed to Edison Liu. Courtesy of A*STAR, http://a-star
.edu.sg/Media/News/PressReleases/tabid/828/art. Accessed December 11, 2009.

of goodwill. The DNA network can become the basis of scenario building in
a cascade of zones constituted by inter-Asia affective interrelations oriented
toward a greater good for the region.

   A napkin doodle attributed to Liu provides a tantalizing glimpse of a possible future for regional science (see figure 7.1). Briefly posted on the website of
A*STAR, it is a rough sketch of trends in the study of family and genetic risks,
presumably in Asia. The spontaneous hand-sketched look suggests a vision
rather than a technical diagram. The center of the sketch is a "zone of solidarity" initiated by the pan-Asian SNP project in which Liu played a key role. The
sketch depicts progressive steps from this zone of scientific collaboration to a
"zone of concern," and then a "zone of action," suggesting a movement from
affective solidarity to practical outcomes, much like the "concept-to-medicine"
transitions in translational research. Then there is the visual exclamation of
"Uma!" The name of the Hindu goddess Parvati, Uma is a dense metaphor for
intelligence and benevolence, plenty and power. Presumably, these linked virtues are the ultimate goal of the SNP project. The placement of "Mendelians"
at the upper left-hand corner of the graph suggests, perhaps, that Mendelian
classical genetics is at a remove from this large-scaled scenario of affective investments afforded by a population genomics infrastructure. If the doodle is
indeed by Liu, this scenario can be read as situating the role of Biopolis at the

center of scientific cooperation among Asian countries. The image promotes the city-state as not only a critical nexus for managing the flows of DNA data across Asia, but also a cosmopolitan and comparatively neutral key site for engendering and fostering scientific solidarity.

As a biomedical hub, the island seems ideally suited to realize the HUGO Asia goal of promoting the novel sharing of genetic materials across the region. As mentioned above, national rivalry and biosovereignty among big Asian countries have precluded any one of them from undertaking DNA analysis for other sites. There was the fear as well that the collection of Asia-wide samples to build a pan-Asian genetic repository would give scientific control of the new science to a major Asian nation. By contrast, tiny Singapore has no Asia-dominating ambitions, at least not of the old kind of throwing one's national heft around. Rather, it wished to be a global hub, the switching station for scientific actors, standards, and traffic across Asia. For instance, compared to India and China, Singapore is interested in the use of genetic data for "the common goal to generate knowledge for clinical or medical applications," as Dr. Lee explained.

Critically, for reputable science, Singapore has locational and infrastructural advantages when compared to the Asian megastates. Given a government system famous for its relative lack of corruption, efficient management practices, and English fluency, Liu impressed upon me, Singapore was "profoundly intelligent" about moving across sectors, investing in knowledge skills, and keeping operations honest. Despite its size, the island is a "global nation that pulls people from everywhere, helping to jump-start scientific talents and calling them our own." The expectation is that rising numbers of foreign residents are expected to form half of the total population of Singapore (which is projected to be about seven million by 2030).[27] The rise of Singapore as a global science hub was in part dependent on its capacity to initiate research links with diverse Asian nations, as exemplified by the SNP network, as well as to enroll not only "world-class" expertise but also its distinctive complement, biological resources from throughout Asia.

Besides, for scientists in many locations drawn together by genomic science, there are concrete gains that come with collaborating with Biopolis. Dr. Ma, the stem cell expert, said carefully, "For any scientist thinking about collaborations, there is the need to weigh what you get." Chinese scientists with world-class training in the West are fine as collaborators, but it is difficult to engage China-trained scientists. Successful research collaborations between Singapore and mainland China also come from having a "good exchange," or a win-win relationship. In return for giving Singaporean scientists access to

People's Republic of China (PRC) research data and samples, researchers in China gain from the "external expertise and credentials" of Biopolis experts. Critically, Singaporean researchers help their mainland Chinese counterparts overcome the English barrier ("coming down soon, though"). Indeed, multi-sited collaborations facilitate joint publications (often with Singaporean scientists as principal investigators), thus helping to raise the international profiles of researchers in China. A metrics-based study of "research facilitation" in the Pan-Asian SNP Consortium from 2004 to 2011 found increasing numbers of coauthored publications. During this period, participation in the research network by mainly forty-four researchers has raised China's publication rate by more than 150 percent.[28] Thus, somewhat reluctantly, Ma admitted that "there is also the ethnic identity issue, and the field of interactions and networks that work to the benefit of Singaporean scientists, I suppose." Here, once again, the issues of ethnicity and cultural understanding seem salient in Singaporeans' capacity to enroll cross-border collaborations, with China, India, Japan, Thailand, and Vietnam, but less so with Muslim-dominated countries, especially Malaysia. Researchers at Biopolis often express regret that, given long-simmering racial tensions and mistrust between the two close neighbors, political mistrust made it hard to do genomic research with their Malaysian counterparts.

Brand-New Ancients

Cosmopolitan science, Steven Shapin discovers in his study of American scientists, reveals how the radical uncertainties of late modern science have unexpectedly deep historical roots.[29] HUGO Asia, a radically novel organization in the region, reaches deep into the murky genomic depths to map a story of brand-new Asian ancients who straddle a specific space-time continuum. The DNA-driven story is haunted by the nightmare of racial backwardness in Asia implied in Western theories of paleoanthropology and bioarchaeology, and infused by dreams of an Asian scientific prowess that may be attained by circulating stories of biological roots and routes. Besides, the explicit association of DNA variation with race carries dense affects about sharing a migratory history, a common biological ancestry, and a set of mutations among peoples scattered in different nations. This first mapping of DNA in Asia has thus managed to make genomic science important, legitimate, and sexy, firing a new kind of self-knowledge that has potential ethical value for the region. The claim of a single migratory swoop of modern humans into the continent challenges a conventional anthropological picture of bifurcated arrival. Marginalized and native populations were drawn into the making of novel knowledge, which showed their hereditary links to larger populations. By integrating indigenous groups in the constitution of majority populations, the SNP project fills in vital gaps in the regional and global picture of human biological and cultural diversity.

The most important role of the Asia-wide SNP consortium, I argue, is as a preparation technology, in infrastructural and discursive forms, to make the genetic past known in order to begin planning for a shared perilous future. The recovery of a new old migratory narration of prehistoric Asian unity is in this sense relentlessly future-oriented, in that it aims to create the affective conditions in which a regional scientific infrastructure, organized toward anticipated but unknown threats to come, might arise. The single-wave theory allows researchers not only to track "what we have been" in our genetic past but also to powerfully reconceptualize "who we are, or can be," as one continent, and perhaps one science in the near future.

Coda

As an ethnographer of Biopolis-driven genomic science, what kinds of anthropological narratives are adequate to the multiscalar, polytemporal, empirically grounded, and conceptual innovative research presented here? As someone who is not just any anthropologist, but who grew up racialized as Chinese in Malaysia, itself mainly a product of British colonialism (with its universalizing set of orientalist fantasies and human sciences), I am both elated that Asian scientists are taking control of the "science of Asia" and unsettled by their strategic manipulations of our immeasurable pasts in order to measure out a possible unified future. Population geneticists associated with HUGO Asia are themselves subjects of mixed cultural ancestries, far-flung places, and cosmopolitan education. They draw on Western DNA science and linguistic and cultural typologies as well as Asian ethnoracial beliefs and politics to systematically reduce the extraordinarily teeming, multiply connected, and heterogeneous peoples (the subjects of "anthropology") into a sterilely inclusive scientific diagram of "Asia." Perhaps a rear-view mirror scenario of civilizational origins is the only technique that can fuel an urgency that transcends a continent of hypernationalism. The scientific branding of Asians as ancients sharing civilizational ideals may yet inspire a coming together as one people to shape a brave new world. In this quest, perhaps scientists in Asia are not that different from their counterparts elsewhere who invoke myths to nurture science, and hope that science in turn will nourish myths.

EIGHT

"VIRUSES DON'T CARRY PASSPORTS"

By the 1970s, Singapore became the cleanest,
safest, prettiest city in the world, but one that lived
in a sea of diseases from other countries.
—DR. WALLACE, 2010

The Return of Tropical Diseases

Deeply tanned, heavily bearded, and wearing a faded Hawaiian shirt, Dr. Wallace stood out among his usually pale, neatly groomed Asian colleagues in shirtsleeves. An American epidemiologist of tropical diseases, Wallace was on a mission. From his early years at the Centers for Disease Control (CDC) in Atlanta, Wallace found that its global sights were set firmly on Africa. But, he confided, there were signs as early as the 1970s that infectious diseases spreading and affecting the global economy were mainly from Asia. So when the call came from a newly established joint Duke Graduate Medical School linked to the National University of Singapore (NUS), he eagerly took up the challenge to "build a CDC-style epidemiology program in Asia."

In colonial Singapore, tropical medicine was established in Middleton Hospital by British Army experts who focused on classic infectious diseases—tuberculosis, pneumonia, typhoid, smallpox—that were the main killers in the swampy and fetid city-port. After independence from the British in 1959, Singapore continued to experience an epidemic of a variety of diseases, including dengue fever. In response, the new national government developed a program to control mosquitoes that, in Wallace's words, "was best in the world." By 1973, the incidence of dengue was down to 5 percent. The control of disease was a key component in the state-led development of the tropical island into a super-clean "air-conditioned nation"[1] in its relentless fight against chaos, dirt, and heat.

In the United States, President Nixon's "war on cancer," beginning in 1971, led to a decline in the focus on infectious diseases (IDs) as the American public began to experience rising rates of chronic diseases—such as heart disease, cancers, and obesity—associated with rich nations.[2] Similarly, in Singapore, the efficient and clean city also shifted its attention to those chronic diseases that were mirror images of those in the United States and for which treatments were more lucrative than for infectious diseases. For decades, the Ministry of Health and local hospitals stopped keeping abreast of tropical diseases, even malaria. Only the severe acute respiratory syndrome (SARS) outbreak in 2003 jolted the public into remembering the ever-present peril of epidemics in the tropics.

In recounting this frightening event, Dr. Lau, a leading Singaporean epidemiologist, remarked, "In peacetime the focus is on modern diseases, but there is the need to prepare for war." The war imagery came to dominate both cancers (see chapters 2 and 3) and infectious diseases, especially newly emerging ones. A new disease center was quickly formed through the amalgamation of Middleton and Tan Tock Seng, the main public hospital on the island. For the first time, Singapore forged a new relationship with the World Health Organization (WHO) and the CDC in Atlanta. The WHO acted as the conduit of information, while Americans contributed funds for the study of newly emerging infectious diseases (NEIDs; examples include SARS, avian flu, the Nipah virus, etc.). In 2005, with the establishment of the Duke-NUS Graduate Medical School, a sophisticated biomedical program came into focus in the fight against tropical Asian diseases.

It is important to recognize that the U.S. military has, since World War II, developed its own research on infectious diseases in Southeast Asia. The Armed Forces Research Institute of the Medical Sciences (AFRIMS) is one of the largest parts of a global network of U.S. Department of Defense overseas medical research laboratories, located in Southeast Asia, the Middle East, and South America. The U.S. Naval Medical Research Unit-2, building upon a legacy of nearly seventy years of U.S. naval research in the Pacific, is now based in Bangkok, and the CDC has a lab there as well. The research unit's focus is on basic and applied research for the development of diagnostic tests, drugs, and vaccines for infectious diseases that threaten military personnel or public health. On the heels of SARS, another American naval medical unit, named NMRC-A, shifted from Jakarta to Singapore. It opens a new chapter in a long and storied line of foreign medical research institutions in Asia. With the U.S. defense medicine system as a fallback, Singapore has situated itself in a constellation of institutions (the Duke-NUS school, Biopolis, and hospitals) poised to take on emerging infectious diseases (EIDs) that are increasingly likely to menace the region and the world at large.

This chapter explores how Singapore's strategies of research, intervention, and preparation for biothreats show critical relationalities and distinctions between the modeling of crises in "natural systems" and in "technical systems." The new anthropological focus on "biosentinels" informs one kind of modeling of infectious disease threats.[3] The technical interventions on the ground are about keeping abreast of threatening objects and events in space and time. The Singaporean lesson from SARS reveals the necessity of bringing a diverse set of techniques for controlling fluid objects and fluid spaces, or vectors and cascading effects related to ever-shifting targets. "Vector" can refer to the course or direction of infectious diseases as well as to international travelers who carry viral threats. In technical systems, "cascade" refers to the layering of events and observations that shape decisions and orient action.[4] My approach emphasizes how an assemblage of interventions, dealing simultaneously with vectors and cascades, operates at different scales from the molecular to the regional and global. Anti-ID strategies from the lab to the airport must constantly prepare to combat the combinability and evolutionary potential of disease vectors. The urgency is global, as epidemics from the tropics can easily spill over into other world regions.

The outline is as follows. The intersection between "multispecies ethnography" and the anthropology of preparedness has opened up a space for anthropologists to be at the forefront of disaster studies. The fluid complexity of actual anti-ID interventions on the ground and in the air involves a constellation of experts and institutions, both situated and global. First, I illuminate how the technical identification of dangerous fluid objects—changeable viruses and their animal-human carriers—or what I call "mutable mobiles"—induces a cascade of spaces to be intervened upon. Instead of the epidemiological "hot spot" as a stable object, in practice, researchers, epidemiologists, and state authorities approach the zoonotic as a series of zones, thereby folding scales—microbial, animal, human, urban, and geographical—of intervention necessitated by the dynamic vectors of deadly diseases through space and time.

Lessons from SARS have primed Singaporean experts to be prepared for cascading events and surprises—especially the combinability and evolvability of viruses and epidemics—that slip easily beyond prior preparations. The SARS experience also highlights Singapore's contribution to global biosecurity, by using its international airport as a spigot to cut off flows of pathogens from tropical Asia. In addition, the mobilization and combination of institutions—including the WHO, global drug companies, and the U.S. military units—is for now necessary to manage pandemics that menace the world.

Biosentinels and Hot Spots

Hot on the heels of scientists, authors and Hollywood producers are predicting that infectious diseases are the global threats of the new century. Bestselling books—*The Great Influenza*, *The Coming Plague*, *The Hot Zone*, *The Viral Storm*, and *Spillover*, to name a few—deliver disturbing information that the world is approaching a time of unprecedented biological threats.[5] Movie houses are packed for films such as *Contagion* and *World War Z*, which, interestingly, focus less on the scientific race to understand viral behavior and more on how fragile the bonds of civilization are during major outbreaks. The latent threat of epidemic circulates through public and popular culture, staging viral menaces as an opportunity to depict the uncertainty of our immediate future, in a kind of epidemiological *Heart of Darkness*. The most critical dangers are the unknown—that is, unrecognized but still somewhat known bugs, so-called NEIDs[6] in the Asian tropics or those EIDs more frequently invoked elsewhere. Preventing a pandemic is an elusive goal that goes beyond lab work and field operations to include the management of viral transmissibility and traffic through human populations.

Certain anthropologists have begun to conduct the ethnography of biotechnical life-support systems and biosecurity threats. A subfield called multispecies ethnography[7] focuses on human–nonhuman exchanges and mutuality as life forms. The multispecies perspective is not an ontological condition but a scientific construction that relates natural and human kinds as cultural complexes. Beyond pointing to human destruction of the environment and self-imperilment, these reframings of human–animal entanglements and atmospheric criticality in the Anthropocene[8] have direct implications for living with and managing zoonosis.

Anthropologists of biosecurity have been attentive to how the modern state can "problematize" issues of security by making "maps of vulnerabilities" that subdivide or overlap with national territory as preparation for events that are potentially catastrophic.[9] More recently, there has been an effort to connect "bioindicators" to "the ecological quality of an ecosystem based on the structure and variety of its populations."[10] Predictive tools for a reemergence of SARS, Frederic Keck argues, could treat "Asian populations as human sentinels . . . or Asia as a sentinel hotspot for the world."[11] This oscillation from national to transnational mapping, from variety to singularity, suggests that mappings and biosentinels are actual management strategies on the ground. As Jerome Whitington has pointed out, biosentinels and hot spots are simulations of potential actions and spaces, and as models, cannot specify the details or experiences of actual "model events."[12] Others as well urge more attention be paid to "material proximities—between animals, humans and objects—that constitute the hotspot" in studying viral management and control.[13]

In other words, strategies of problematization generate different spaces of intervention at multiple scales. Below, I argue that, in Singapore, the shifting vectors of diseases engender a cascade of data that inform and identify different spaces of problem solving. In other words, mutable viruses and mutating spaces of contagion demand political and scientific interventions that are flexible and scalable in order to manage the risk potentiality and risk management of viruses in motion.

Fluid Objects, Cascading Scales

Technologies for dealing with potential biothreats, I argue, rely on the management of fluid objects and fluid spaces. On the one hand, there are the fluid objects that by being changeable in transit become threatening to life. Alongside Bruno Latour's notion of immutable mobiles, I call these fluid, mutating objects "mutable mobiles." Viruses and infected animals and humans are all fluid objects that are mutable because they flow across different ecological niches, political spaces, and temporal horizons. On the other hand, mutable mobiles such as dangerous pathogens or virus-carrying people move through a series of spaces from wild jungle spots to urban centers to global transportation hubs. Death-dealing viruses mutate through time, and their lethalness lies in their capacities to change at a rapid rate and evolve into more threatening microbes. Concurrently, infected patients can spread the disease as they travel through ecological and political spaces, thus geographically enhancing the super-spreader effect. Viral management strategies using bioindicators would need to track and check the mobility of pathogens and nonhuman and human carriers in order to stem the spread of diseases beyond sites of emergence.

The accumulation of indicators and events of viral spread therefore engenders a cascade of scales for intervention. Microbiologists use "cascade" to describe the chain of signaling that constitutes the immune response of the organism to an invading pathogen. Similarly, researchers, health workers, and officials are exquisitely aware of the spatial fluidity associated with moving viruses and the need to divide up spaces in order to contain the spread of contagions. Michel Foucault has argued that medical classification, surveillance, and containment have historically generated a logic of "tertiary spatialization" to segregate and contain the spread of disease carriers.[14] It is in this sense that spatialities are enacted not only by the movements of microbes and people, but also by techniques and knowledge of science and governing.[15]

Contemporary preparation technologies for emerging contagions, I argue, combine "figures of warning" and a welter of spatializing practices, as the medical surveillance of virus mutability is coordinated with strategies to spatially contain or eliminate vectors of deadly diseases. Apparatuses of security "have the constant tendency to expand; they are centrifugal."[16] Security measures, I argue, involve cascading effects as well as the folding of multiple scales into an assemblage of intervention, which reworks borders between both human and nonhuman identifications with spaces of safety and of threat. Post-SARS, Singapore's approach aims to coordinate the work of cell biologists, health experts, and public health workers such that interventions happen simultaneously from the microbial to the social and the regional scales.

Rezoning the Zoonotic

In the post-AIDS and post-SARS world, disease scientists have been anticipating new global pandemics. They agree that "if there *is* a Next Big One, it will be zoonotic."[17] "Zoonotic" refers to any animal infection that spills over to the human species, resulting in often lethal infectious diseases such as SARS, AIDS, Ebola, and Middle East respiratory syndrome (MERS). Given the relentless fragmentation of ecosystems and rapid rates of urbanization across the tropical world, pathogens that used to dwell within isolated organisms are leaping from animals to humans at a frighteningly fast rate. The proliferation of dangerous bugs necessitates recon work, figuring human–animal (species) boundaries as sites of potent medical and political investment. For instance, shifts in disease vectors, environmental conditions, and rates of transmission are recasting South China (a source of the coronavirus for SARS, and mutated viruses for avian flu) and Southeast Asia (a source of the Nipah virus) not only as regions of human–pathogen interactions, but also as critical zones through which infectious diseases can be disseminated to the world at large.

Modern Asian nations have repeatedly deployed "zoning technologies," from establishing special economic zones for linking up with global capital circuits, to security areas for dealing with refugees and irredentist movements.[18] The indelible impact of SARS in 1998 in Southern China that rapidly disseminated into Southeast Asia has overlain the image of a booming political economic region with one that is a sieve of leaking viral pathogens to proliferate through the whole world. SARS intensified the vulnerability of the region as a network of sites through which the deadly viruses circulate, inspiring a sense of hypersecurity and the need to come together on fast-moving diseases, including avian flu.[19] Cascading health scares have prompted a re-narration of "development" as an engine of epidemiological change, taking new relations between viruses and species as an unexpected modern product and political object.

For the above reasons, technologies of preparation and intervention do not deal with a single global hot spot, but instead generate multiple zones at different scales. Observations of mobile species-assemblages produce micro scales that differentiate between viral pathogens and reservoir animals, between infected animals and human beings, and between healthy people and those who are infected. At the same time, political authorities become alert to shifting spaces of human safety and infection that require different zonal treatments. From the Singapore vantage point, biosecurity for the region requires the coordination of micro-zoning for species-assemblages with a macro-zoning of human border crossings. Disease surveillance, study, and viral management in combination generate cascading effects that recast the notion of "Asia" as a space of multiscale flows of pathogens leaping among species and countries.

In the chapter's opening quote, Wallace articulated a regional topology when he called Singapore "the cleanest, safest, prettiest city in the world, but one that lived in a sea of diseases." From the epidemiological viewpoint, the island's location is connected by its very propinquity to other Asian sites in a network of viral flow and infection transmission. "What is exotic to Western medicine is at home here," Wallace observed from his vantage point one degree north of the equator. For researchers based in Singapore, species-jumping infections bind nonhuman and human organisms in discrete relations within the Asian body and its geography as well. The Asian tropics and subtropics contain a rich and ever-changing brew of dangerous viruses that could rapidly combine with animals and humans and spread across the world. Besides disease ecology, high rates of deforestation and urbanization have accompanied unrelenting economic development over the past three decades, contributing to the definition of Asia as a hot spot for global pandemics.

There are multiple zoonotic layers linking nonhuman and human organisms in spreading lethal infections. South China and Southeast Asia are the epicenter of influenza and of potentially fatal respiratory illnesses such as SARS. This disease ecology, which gives rise to viruses with pandemic potential, has been shaped by a combination of agricultural systems and accelerated economic transformation. Asian farming practices frequently combine different species (waterfowl, pigs) that are the favored hosts of the flu virus.[20] For instance, the SARS virus was found to have emerged from horseshoe bats in China, but it was carried by other species, such as civet cats, that acted as amplifier hosts in Guangdong's wildlife markets, transmitting the pathogens to human beings.[21] In Malaysia, pig farms exposed to wild bats seeking new habitats became the source of the Nipah virus. The unprecedented disruptions of natural and human habitats as well as the effects of climate change are reconfiguring the disease ecology, producing novel conditions that speed up the release of lethal pathogens into people in the transformed Asian ecosystem.

The genomic analysis of viral animal–human transmissibility identifies the zoonotic as a series of jumps from deadly pathogens to pathogens and animals to animals and humans and to humans and humans. This tracking captures the contingent viral spiral within a zoonotic ecosystem. For instance, bugs can be carried inconspicuously on a number of animal hosts before leaping to human beings, who become the reservoirs for human transmission. While the zoonotic spillover is a critical relationship that tightens conceptual connections among microbes, animals, humans, and the ecosystem, governing strategies discern different scales of effects and a need for graduated governance.[22]

The outbreak of SARS emphasized the zoonotic within the human, thus coming to define a new dimension of being "Asian." NEIDs mainly attack Asian peoples because they live and interact within the same fast-changing environment. Other potential pandemic diseases include H_5N_1 influenza and, increasingly, the recent H_7N_9, which has a much higher rate of fatality (over 50 percent) than SARS has. The pervasive presence of known and anticipated infectious diseases preying on people has spurred genetic investigations into how dangerous bugs and human hosts are connected. If diseases define new spatialities of zones and buffers, these necessarily recode human bodies into the determinants of an important spatiality. Biomedical practices thus work through a kind of mesh that connects identifiable zones from the cellular scale up.

Folding Scales of Intervention

As in other areas of Singapore genomic science, the circulating but locatable markers of humans, animals, and microbes as "Asian" play a big role in folding together different scales of intervention, whether in the lab, the city, or the world.

THE VIRAL-ETHNIC FOLD

In recent years, Singapore-based experiments have produced remarkable advances in the diagnosis and management of dengue fever. Current projects at A*STAR and its partnering institutions include not only controlling the spread of the *Aedes aegypti* mosquito, the major vector of dengue, but also research into disease genomics and clinical studies. In the lab, the molecular approach allows for intervening at the point of interaction between the pathogen genome and the host genome. At a Duke-NUS lab, genome sequencing of RNA viruses (that cause SARS, influenza, and hepatitis C, or the major infectious diseases in the region) helps scientists to construct phylogenetic trees (models of evolutionary relationships) that point to their originating sites, that is, in Southeast Asia. Furthermore, the study of variability in human responses to pathogens further localizes infectious diseases as inexorably linked to Asia and its inhabitants. Biopolis partnerships with drug firms have developed kits for detecting SARS from blood tests that can be carried out in hospitals. Other public-private projects design tools for screening influenza virus mutations and an H1N1 vaccine. While the medical devices and medicines can benefit the world, they are firmly oriented toward human and nonhuman life forms in Asia, by recognizing the immunological variability in human responses to infectious diseases.

In other words, molecular investigation of viral-human interactions supports that some populations are asymptomatic to some viruses and others are not. Wallace explained that in host genomics the aim is to identify the key host pathways involved in infection through the use of whole genome expression arrays. The finding of genetic variations in patient immune responses to infectious diseases (dengue, tuberculosis, sepsis or blood infection, etc.) permits a kind of patient profiling. For instance, dengue is a menacing infectious disease that infects close to one hundred million people each year, most of them in Asia. In severe cases of the so-called breakbone fever (symptoms include joint pain), the mortality rate is 30 percent. Dr. Pan at the Duke NEID program noted that while SARS receives more global attention, dengue fever was "an ever present danger to populations in the region." He said that the data on host-pathogen interactions show that Chinese subjects have the highest incidence of all ethnic groups in the region. Association studies show that there is a connection between Asian genetic variants and particular responses to infectious diseases, for instance, among Asian subjects, the responses to dengue fever are 80 percent asymptomatic for a few days. Given its widespread resurgence today, the first genomewide association study for dengue fever was recently conducted by the Genome Institute, together with Wellcome Trust's and Oxford University's research projects in Vietnam. By studying the genomes of children with severe dengue, the team identified genes that may increase a child's susceptibility to dengue shock syndrome, a life-threatening form of neurological complication.[23] In the midst of novel lab experiments—the study of human-viral interaction, viral genome evolution, and human immunology—Singapore authorities also revive their earlier antidengue campaigns.

URBAN SURVEILLANCE

In the 1960s, dengue epidemics were almost a yearly occurrence, until a Singapore national *Aedes* control program, combined with health education and law enforcement, brought down disease rates from 40 percent to under 10 percent per hundred thousand. In recent years, accelerated deforestation in Singapore and throughout Southeast Asia has made the urban environment a reservoir for the flourishing of the mosquito-borne dengue fever.

Disease surveillance requires the differentiation between spaces of spillover and of sanitation. In contemporary Singapore, the antidengue campaign deploys an "active cluster" approach to identify localities with active transmission where intervention is targeted. To alert residents, public health efforts, which included a color-coded map for different levels of urban infestation, focused on the eradication of potential spaces for the transfer of diseases. The National Environmental Agency has adopted a multipronged approach to control dengue by denying the *Aedes* mosquito places to breed (source reduction). The main thrusts include preventive surveillance and control, public education, and community involvement and research.[24] At the same time, community involvement is rigorously enforced. Homeowners found to have places on their properties with breeding places for mosquitoes were fined, and repeat offenders faced the threat of jail. Disinfection exercises were a regular part of the urban battle against the mosquito, including weekly rounds by "fogging" machines to spray repellent on Singapore's parklike surroundings. In May 2010, I was able to have a drink in a garden terrace that was surprisingly free of buzzing insects. Infectious disease pest controls are thus entangled with the image of the city as well managed and authoritarian in enforcing controls over spaces of potential contagion. Thus the sanitized Asian city became a surveillance hub for NEIDs in the tropics.

At the turn of the century, another kind of clustering emerged to coordinate the management of disease vectors from Singapore. The WHO and the CDC, as well as Duke University Medical School, chose Singapore to be a center for tracking, diagnosing, and coordinating, as well as assisting in, the development of vaccines for new infectious diseases.[25] The medical school complements the British-style medical model by training American-type physician-researchers, who are exposed to clinically related research, thereby increasing their biomedical expertise to meet future challenges. Surveillance and research on infectious diseases are conducted by Duke-NUS and the entire network of public hospitals and medical institutes. The city-state has a combination of biomedical and political effectiveness: disease medicine and disease surveillance, political controls of movement, and access to medical expertise and drugs. This mix of scientific and political capacities, perhaps unique in the region, lays the groundwork for the development of situated expertise similar to a CDC-style center for the rest of Southeast Asia.

THE MOSQUITO TRAP

Then there is the rescaling of antidengue efforts in experiments at the level of viral evolution and host immunology. The recently formed Singapore Immunology Network (SIgN), in alliance with the Singapore Novartis Institute for Tropical Diseases and the Beijing Institute of Microbiology and Epidemiology, is experimenting to disable the molecular workings of the dengue virus. The viral RNA is capable of "imitating" the RNA of the host cells, to the extent that the virus can remain undetected and evade the host's normal defense systems. A novel strategy invented by Dr. Katja Fink of SIgN introduces a weakened mutant dengue virus that triggers a strong protective response from the host's immune system. Dr. Fink explains, "We assumed that viruses lacking this 'hiding strategy' would trigger a robust immune response that would stop them in the early stages of infection, a requirement of live attenuated viruses used as vaccines."[26] This means that by triggering the *Aedes* mosquito's immune system, the mutant dengue virus cannot infect the insect, thus potentially ending the mosquito's role as a super spreader of the frequently fatal disease.[27] The new immunology technique holds the potential for a major breakthrough in dengue vaccine development.

Such innovations in infectious disease medicines can potentially ensure that Singapore and the surrounding region will have an independent supply of vaccines in the near future. By zeroing in on the genetic interplay between dangerous bugs and Asian populations, researchers in Singapore define the objects that constitute the regional space of their disease science. The zoonotic zone in the Asian body identifies its susceptibility to infectious diseases. At the same time, the demographic, cast as molecular, genomic, and epigenetic susceptibilities, also defines a space in the interaction of peoples and bugs and diseases.

The folding of different interventions—microbial, animal, human, geographical, scientific, and epidemiological—in Singapore also enacts different reiterations of the zoonotic as multiple zones within and among viruses, humans, and ecologies in Asia. The historically unprecedented fragmentation of wild places, combined with farming practices and the urban reservoirs of infectious diseases, creates conditions for pathogens to interact with human hosts and to start pandemics. There is thus a double ethnic heuristic and technical appropriation of "ethnic places" in play, making the zoonotic within humans and between humans in the ecosystem a part of predictive medicine.

But Singapore as surveillance hub is also about scanning the horizon for potential doomsday diseases. Besides managing the reemergence of "old" infectious diseases—dengue hemorrhagic fever, avian influenza, Japanese encephalitis—researchers and policy makers focus critical attention on negotiating the shifting borders between pathogens and their human victims: borders that maximize the potentials for the next nightmarish outbreak.

Lessons from SARS: Evolvability and Combinability

Biosciences, by definition, deal with mutating objects and spaces, with the emergent and the still unknowable in space and time. Dangerous bugs are mutable mobiles par excellence, objects apprehended by science that are in constant motion in form, time, and space, so that critical preparation techniques must not only track, but also anticipate their mutability and combinability.

Within the seething tropics, microbes have a high degree of genetic mutation and tendency toward "reassortment" (of genes from related strains), which make them highly adaptable to changing conditions within new hosts, giving them "intrinsic evolvability."[28] Local researchers are thus extremely attentive to the rapid evolvability or recombination of dangerous microbes, as well as their interactions with animal and human hosts in environments that are also impacted by climate change. Building knowledge on a spiral of interactions that connects viruses to human bodies and places, researchers are acutely aware of the different temporalities and spaces of emerging pathogenic threats.

As the source of many viral infections menacing human populations, the bat has become a target of comparative genomic analysis. At Duke-NUS, Dr. Wang Linfa, the leader of the infectious diseases program, is an expert on bat genomics. He discovered the bats' ability to carry a large number of deadly viruses that do not directly infect them but that, in disrupted environments, can infect human beings. The timeline connecting released viruses, reservoir animals, and humans can be traced at the genomic level.[29] Scientists, however, are never prepared enough for NEIDs in the tropics. Even recently discovered pathogens—SARS, Marburg, Hendra, Nipah, Menangle viruses—are being succeeded, every few months, by unrecognized deadly bugs that are being released from disrupted environments to threaten human beings. In short, the mutable mobile virus has become a new political object that threatens to break through multiple borders of containment, natural as well as political.

Governing the near future thus demands new conceptions of fluid objects and the spaces they infiltrate or are kept out of. After all, dangerous bugs can hitchhike on air travelers to all corners of the world. A twenty-four-hour spread of infectious diseases globally can ignite fear and chaos in regions far from the originating site. The situation is exacerbated by the fact that many deadly diseases such as SARS are asymptomatic but still infective for a day or so, thus making them difficult to detect and control. In an era of pandemics spreading at the speed of jet planes, planning for potential pandemics must grapple with often invisible air-borne mutable objects that piggyback on human travelers as well as the question of how to regulate flows of disease-carrying subjects (super spreaders) to other places. There is no stable or uniform sentinel spot or space for these mutable mobiles. Therefore, technologies that anticipate potential future biothreats must calculate infection temporalities and travel spatialities in order to be able to impose sanitary controls on bodies and to delink spaces of infection from buffer zones.

The SARS outbreak, which caught China totally off-guard, rudely alerted Asian governments to the risks of virulent infectious diseases that spread rapidly because of air travel.[30] Singapore began to view populations as not only a source of skills but also a conduit of diseases. The increasing flows of people in and through the island not only marked it as a global transfer point, but as a vector of potential biothreats. By the next decade, medical facilities and expertise have become increasingly central to the self-definition of

Singapore as a critical site for developing techniques to confront anticipated future biothreats and for priming citizens for living in an epidemic-prone region. At stake as well is the threat that pandemics posed to modern developmental and living standards recently achieved in the region.

Because SARS spread like wildfire, Singapore health experts became acutely sensitive to the need for barriers between viral pathogens and humans, humans and humans, and patients and doctors. SARS is asymptomatic but infective for a few days, making the disease hard to detect and control before it has potentially spread through more asymptomatic hosts. Infected people, including attending health workers, are killed within a matter of weeks. The first line of defense therefore is to screen out the pathogens carried by air travelers. Soon after the SARS epidemic, I arrived at Singapore's Changi airport and was confronted by heat-sensing machines, which monitored visitors for exposure to infections, especially SARS.[31] Since that perilous time, airports throughout Asia have installed devices to screen for infectious travelers and, less systematically, one suspects, for human terrorists. Other Singaporean institutions learned costly lessons about the need to manage temporal and spatial flows of viruses and their human carriers.

During the SARS outbreak in Singapore, "for the first two weeks, we thought it was the end of the world," recalled an American geneticist then at Biopolis. But the city-state was more ready to face biomedical risks and challenges than other Asian sites where SARS spread rapidly and interventions were slow or spotty. Pan, currently in the Duke EID program, remembered the alarming days of the SARS outbreak (March to May 2003) on the island. "Every lab in Singapore worked on SARS. My lab dropped its projects and grew SARS material for diagnosis and research work. It was immediately identified as a national need and clear goal, and we have the expertise to help." The island already had in place a series of excellent health institutions to respond efficiently to the crisis. The Ministry of Health coordinated hospitals and research institutes to work together seamlessly for a period of three to four months. Tan Tock Seng Hospital took the lead, and ICUs for SARS patients were set up. Tents erected in hospital compounds screened patients. In the midst of plunging GDP, Pan reported that government servants came together to do "what it takes to control the disease."

Out of the SARS crisis came stories about the crucial roles of doctors and biomedical expertise in a cultural environment that already reveres science. In Singapore, doctors became national heroes for fighting (and dying) in the war against SARS. They resolutely managed the front lines of defense against a terrifying new infectious disease. A physician who led a SARS workgroup noted that the ethical quandaries in managing the SARS crisis are akin to those portrayed by Albert Camus in *The Plague*. An article in the local medical journal urged doctors to "strive their utmost to be healers," to "be resolute and vigilant," because "the pestilence never dies, but returns again to menace the happy city."[32]

Such narratives liken scientists to self-sacrificing heroes who draw the line against potential flows of deadly diseases from elsewhere in the region. In this climate of health vigilance, there was increased focus on the mutability of viruses (SARS, avian flu, etc.) and worries about rapid nonhuman to human transmission in a region densely interlinked by commerce, migration, travel, and environmental vectors. For the first time, economic and health futures become interwoven in algorithms of risk and uncertainty. Whereas citizens were previously uninterested in the Biopolis project, after SARS they began to warm up to the idea that Singapore needed to develop the scientific expertise to keep abreast of mutability in viruses and in patients, and they realized that the possibility of the reemergence of SARS and encounters with potential others was terrifying.

Global pandemics are shaping a biopolitics that manifests the call to "think globally, act locally." Situated intervention in practice includes the politics of exception when it comes to the management of fluid objects and the spaces they shape. At the 2009 opening ceremony of the Duke-NUS Graduate Medical School, the president of NUS emphasized spatial over biomedical strategies for confronting proliferating diseases. "The global trends of rapid economic development and human population growth suggest that Asia will continue to be the epicenter for the emergence of epidemic infectious diseases that can spread rapidly via air travel. As we are frequently told, viruses don't carry passports." Vigilance against global pandemics, he continued, "cannot be tackled by any single country alone. We need to have close collaboration between countries, and with global organizations, to coordinate surveillance, containment and response to minimize the impact of infectious disease outbreaks."[33]

As a global site of incoming pathogens, Singapore was well positioned to coordinate strategies of "containment and response" by controlling international flows of human (and animal) carriers of dangerous diseases. Because viruses are defined by their capacity to spread and assemble bodies into pathways and vectors, a quarantine becomes a potential spatial strategy for diseases in general.

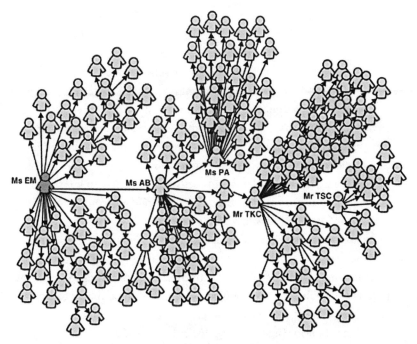

FIG. 8.1. The SARS super-spreader phenomenon—Singapore, 2003. http://www.cdc.gov /mmwr/preview/mmwrhtml/mm5218a1.htm. Courtesy of Centers for Disease Control and Prevention, Atlanta.

The past experience of the government with SARS "super spreaders," or people who have the capacity to infect ten or more other people, emboldened the imposition of a quarantine during the pandemic. In infectious diseases, the super-spreader effect is created by the aerosol spread of the virus by coughing, a form of airborne transmission for other deadly infectious diseases (rubella and laryngeal tuberculosis). In a CDC diagram (see figure 8.1), Singaporean health authorities furnished information on five SARS super-spreader cases, detailing the multiplier rate of transmission.[34] The dramatic spreader effect was exacerbated by the fact that infectivity was hard to detect within the first day or two. Severely ill patients were hidden reservoirs of infection in the hospitals and in the community at large. One infamous case was a super spreader who survived the encounter with SARS but infected more than one hundred people.

Early in the SARS outbreak, initial contacts among patients, their relatives, and attending healers led to dozens of deaths in Singapore hospitals. Very quickly, strict infection-detection practices were adopted, from isolating suspected and probable SARS cases, to careful uses of masks and gloves, hand washing, and daily temperature checks for all health workers. Such clinical practices expanded to become a general template for the management of society at large. In public places such as open-air "wet markets" and schools where a cluster of illnesses was detected, authorities closed down business as usual. People in contact with anyone with SARS were quarantined at home for at least ten days. To prevent secondary infections, suspected SARS carriers were required to wear electronic tracking bracelets. They were closely monitored and had food delivered to the house. Quarantine breakers were fined or jailed. In order to control the circulation of pathogens and super-spreader victims, sensors were placed at transportation hubs, and cab drivers were subjected to checks of their daily temperature. And so the incidence of SARS ended within a few months.

Compared to Hong Kong, China, and Taiwan, where the disease spread rapidly, the CDC in Atlanta noted that "a much more expanded policy for contact tracing and home quarantine has been instituted in Singapore."[35] This toughness and swift action, Pan noted, were due to the political capacity to make and implement difficult decisions as the outbreak unfolded. "We had the confidence to bring drastic actions to control SARS." At least in Singapore, the authoritarian enforcement of quarantine was the most effective way to stop SARS in its tracks. The politics of authoritarianism is a dimension of a kind of ethno-zoonotic regionalization as a strategy for containing potentially virulent air-borne diseases that violate all borders.

The Singaporean experience with SARS is a textbook case of how a good basic health system is the most important infrastructure to end epidemics within local bounds. Ongoing epidemics of dengue fever in Brazil, and Ebola in West Africa, raise the specter of an uncontainable spread overseas. At the point of an epidemic spillover, international coordination, as in the case of Ebola, becomes very contingent and poor. Complex infrastructures—transportation and military systems, for instance—contain risks that can cascade into major catastrophic events, and they must collaborate in order to prepare for the next major pandemic emerging from Asia.

The City as Spigot

In preparing for the potential fearful contagion, political practices proliferate to spatialize the origin and distribution of disease in order to contain it. Above, I mentioned the Singaporean response to dengue and SARS through the creation of multiple spaces of segregation within clinics, hospitals, and the sanitized city. But international fears of contagion add another layer of spatialization. The speedup of infection and contagions, abetted by the network and density of airline travel, has made more urgent the need to enforce a strategic screening of flows of dangerous objects to the rest of the world. For instance, looking back on SARS as the perfect storm of global vulnerability, a CDC official cautioned, "We have no capacity to predict where it's going or how large it is ultimately going to be."[36] In other words, the logics of zoning for surveillance and containment of infectious diseases thus introduce the elasticity of scales in a more global context.

Global coordination seeks to develop capacities in so-called source areas to quickly detect, name, and diagnose health threats in real time. An emerging topology as an infrastructural shield against novel pathogens and contagions finds a potential site in Singapore, if only to manage the inside/outside flows of nonhumans and humans across many levels of international movements. While their language has been one of developing "a CDC-style center," the spatialization of critical skill sets and of movements of disparate fluid objects, both rejected and desired, is a vital aspect of regional surveillance and air-travel controls.

Thus Singapore as an epidemiological spigot is about spreading knowledge as well as containing any emerging pandemic within the region. The city as a clinic discussed above also acts as a super spreader of American scientific "best practices" in epidemiology and intervention. The city mobilizes a spectrum of actors—lab researchers, epidemiologists, fieldworkers, immigrant officers, the police, vaccine suppliers—to develop prevention technologies such as tracking sentinel animals and human populations. Singapore's lead in the training of health personnel in developing Asian countries was prompted in part by the United States setting up the Regional Emergency Disease Initiative (REDI) Center there. The goal of REDI is to act "as a catalyst for regional collaboration on research related to infectious diseases of particular relevance to the Asia-Pacific region."[37] With funds from the United States, health experts from Singapore, Australia, and New Zealand have come together to train public health officials, researchers, clinicians, and other health professionals.

Under Wallace's guidance, the Duke-NUS program is building a viral forecasting system for Southeast Asia by spreading local capabilities of field observations in tropical jungles from which pathogens break loose to threaten human populations. Singaporean scientists have helped establish field labs for collecting data from animal reservoirs by working closely with local health workers in Thailand, Vietnam, Malaysia, Indonesia, Sri Lanka, India, and so on. As Nathan Wolfe has argued, fieldworkers must be trained to respond to "viral chatter" or everyday communications about viruses infecting humans in particular sites, thus signaling a looming plague.[38] Furthermore, they have to be trained to monitor their own diseases and to do their own diagnostic projects potentially as well as the CDC can.

The U.S. involvement in upgrading clinical skills and regulations in developing countries includes other agencies besides the CDC. In Singapore, a naval medical research program that monitors diseases among servicemen can be pulled into action in the event of a health emergency. The AFRIMS, based in Bangkok, which is focused on infectious diseases, also contributes expertise on virology to mainland Southeast and South Asia.[39] The potential effects of U.S.-led initiatives to distribute disease prevention practices, with Singapore as the regional center, are still limited, and such initiatives must circumvent national roadblocks to the configuration of a regional space of biomedical expertise.

Southeast Asian nations tend to assert sovereign power to control "bad" news" (e.g., of epidemics) and are sometimes reluctant to share health data, research, and expenditures. Furthermore, they are, for historical reasons, sensitive to U.S. surveillance of conditions in their countries. Thus, other than the Duke-NUS partnerships with small field projects in some countries, it was not clear how prepared REDI has been in developing intra-Asian health collaborations. Pan explained that building a regional surveillance system would require a formal agreement by the Association of Southeast Asian Nations. Only such a political pact will permit access for transnational academic research, and there must be agreement from each nation to work with the same protocols in a standard way. At the time of my visit, a WHO-led coordination of knowledge sharing about NEIDs was the most significant collaborative scale-up achieved so far.

Perhaps Singapore's desire to be a globally recognized leader of infectious disease research in Asia will come not so much from spreading scientific best practices throughout Southeast Asia, but from its potentiality as a garrison state. Indeed, in the 1970s, when the city-state felt under siege from hostile neighbors, advisors from the paradigmatic garrison state Israel, similarly surrounded by purportedly unfriendly Muslim neighbors, were smuggled in to advise government officials on military security. In contemporary times, this garrison-state model may be the basis of potential interventions into biothreats rather than military ones. For instance, beyond the images of Singapore as a clean city floating in an Asian sea of diseases, and as a transportation hub that receives over fifty million visitors each year who may bear incoming pathogens from neighboring countries, there is also the potential of Singapore as a political-epidemiological spigot that controls the influx of fluid objects and is thus to act as a strategic gatekeeper of buffer and safe zones in times of global health emergencies.

The authoritarian efficiency for which Singapore is famous makes it a potential garrison state against global pandemics. The garrison-state model grew out of countries that considered themselves under siege and thus obsessed with state survival. Indeed, Singapore has long borrowed some practices from the garrison state par excellence, Israel, from its training of elite armed forces to state surveillance. Another leaf from the Israeli playbook for dealing with uncertainty is the pragmatic and high-tech structure in place for enforcing border controls, against the flows of not only infected persons, but also pets and other life forms. SARS was a trigger event for the buildup of surveillance expertise and controls of borders, and it was also an experiment for planning future health emergencies that inevitably cross borders. As a super-efficient city-state with lockdown controls, the city-state at the crossroads of global travel and commerce is developing into a hot spot for crisis management. Singapore-based technical expertise and pragmatic politics combine state security and global biosecurity as a single potential goal in planning for the next big biological catastrophe. Asian authoritarian gatekeeping is in many ways a potential first line of defense against the international spread of pandemics from Asia.

Furthermore, the need to have preparation technologies in place also requires us to rethink the connection between the politics of securitization and biocapitalism. Pan told me that normally most hospitals have thirty-day supplies of important drugs and vaccines. The presence of drug companies is therefore very important to keep supplies delivered speedily during a health crisis. The SARS experience highlighted the need for keeping the supply chain of drugs and vaccines open for potential biothreats. He emphasized that, in a future pandemic, the state should be capable not only of enforcing strict travel rules, but also of letting through essential medicine and foods. Technologies of distributed preparedness, at least in Asian epidemics, anticipate the articulation of the politics of health authoritarianism and the power of big pharma both to draw lines of quarantine and to restore health and security.

The prediction and crisis management of global pandemics in Singapore hugely benefit from the local presence of hundreds of drug companies, including giants such as GlaxoSmithKline, Pfizer, and Merck. Drug companies manufacture drugs as well as develop new therapies for diseases originating in the region. The Novartis Institute for Tropical Diseases in partnership with a Singapore state agency helped develop the new immune strategy to prevent the dengue virus from escaping the defense system of the carrier mosquito. Besides developing scientific capacities in the global south, the institute is developing new medicines to treat long-neglected infectious diseases such as malaria, tuberculosis, and dengue fever. Furthermore, the novel therapies will be sold at cost to poor countries where the diseases are endemic (the Gates Foundation provides funds to help make this possible). Thus, ironically, big pharma companies with offices in global Singapore become part of the chain of supplies both in planning for epidemics and in containing them. In short, the combination of big pharma and authoritarian powers to meet the next big health crisis is more explicitly planned and actionable in Singapore than almost anywhere else in Asia or the developing world.

Conclusion: Biosecurity Assemblage

The expectation of Asia-sourced infectious diseases to globalize rapidly has required a technical reconceptualization of the region as shifting vectors of diseases and cascading events capable of enfolding other places in a spreading biothreat. Singapore has developed different scientific and infrastructural systems that enfold different scales of intervention. While researchers work on the genomic and immunological aspects of dangerous viruses and their hosts, technocrats strive to control the influx and outflow of floating objects (knowledge, dangerous microbes, infected animals and humans, health workers, critical drugs, etc.) in a topology of security. The Singapore case demonstrates complex planning for and anticipation of free-flowing viral pathogens that involves a combination of ideas, techniques, and institutions that scholars often keep apart in their analysis. Whereas bioindicators and hot spots are important models of emerging diseases, other technologies include both a dynamic and heterogeneous approach to disease-infested regions and a recognition of the fluidity and evolvability of target objects and spaces. Just as important, in order to contain the chain of spillovers—from deadly viruses to animals to humans to other places—the state has a vital role to play. Confronting epidemic uncertainty would require an articulation of state security and global security apparatuses, as well as, in Asia's example, the collaboration of strong states and big pharma.

To be a potential biosecurity stronghold, Singapore has developed biomedical research and technical systems capable of patrolling the disease borders of an identified Asian zoonotic world. Instead of the unmaking of government, technologies of preparation and scenario making are fundamental elements for governing the near future in the Asian tropics. Techniques for managing the fluidity of microbes, organisms, and spaces are vital for staying ahead of the moving horizon of potentially deadly diseases that move rapidly to menace peoples in other sites. As the involvement of American health and military agencies shows, the mix of technical expertise, political authoritarianism, and big pharma is forming an emerging topology of biosecurity against potential global contagions.

The 2014 Ebola outbreak in West Africa provides a vivid demonstration of cascading events in the relative absence of basic health systems and international coordination. Ebola became a runaway epidemic that claimed thousands of lives over a six-month period before governments swung into action. In a highly mobile region with weak health systems, the lack of information and spatial controls fueled an alarming spread of the pandemic, sending shockwaves from Guinea to many African countries. Vinh-Kim Nguyen observes that Africa can be thought of as a "republic of therapy" woefully dependent on a multitude of nongovernmental organizations and global health entities such as the Gates Foundation for meeting the threats of infectious diseases.[40] The Ebola epidemic sparked resistances to hospitals (viewed as sites of contagion) and quarantines, as troops were slowly deployed to protect and secure lives. Weak public health systems and limited hospital beds contributed to the spread of the disease because of the home treatment of patients. Much too slowly, international help trickled in to mitigate what turned out to be a long-term epidemic. Cuba sent the largest medical team to Sierra Leone, while other countries sent money rather than doctors. A potential drug for Ebola developed at Emory University was used on infected American health workers, raising questions about the economics of drug development and distribution for people in many developing countries where the disease had erupted or will erupt.

But the nature of epidemics ruptures borders, and technologies of uncertainty require complex coordinations of international expertise and infrastructures to manage mobile pathogens, animals, and patients and to keep them under quarantine conditions. Technologies of preparation identify a series of scales—from the molecular to the corporeal, social, political, and global—that shapes the topology of global biosecurity.

Many places in the world lack basic health systems, the first line of defense. It therefore falls on relatively affluent and well-organized places such as Singapore to develop the ambition to be a regional hub of health sustainability. Given its location, the city-state's potential vulnerability to many intersecting traffics also creates its potential for controlling flows of deadly diseases. By understanding the relations and places that Singapore gathers as a hub, the airport needs to act as a political spigot controlling different flows in the event of emergencies. In Israel, the state garrisons itself against its purportedly unfriendly Muslim neighbors as well as potential pandemics. In Singapore, the state as spigot can operate most explicitly as a cordon for the world against "Asia" wrought as a disease-ridden place (but also as a source of potential terrorists). A quarantine against deadly infectious diseases interacts with military-political barriers against the body politic, writ large. While radical uncertainty remains, such preparations are among the best to prevent the leaking of tropical diseases into the rest of the world.

NINE

THE "ATHLETE GENE" IN CHINA'S FUTURE

> Life is of sequence—A.G.C.T.
> Life is digital
> Information is the language of life
> Not the analog.

In his epigrammatic style, Dr. Henry Yang Huangming, the founder and then president of BGI Genomics, China, set out his "philosophy" of "why life is life" at a 2010 Harvard-sponsored conference about Asia's vision in the twenty-first century. Yang was introducing BGI Genomics (henceforth BGI) to the rest of Asia, describing it as "the largest genome sequencing center in the world." The institution aims to sequence the genomes of any and every living thing: humans, animals, plants, fungi, and microbes. On the basis of increasing the ease in mapping the long lists of nucleotides and combinations of DNA markers for individual organisms, Yang declared, to gathering unease in the crowd, "Bioscience and bioeconomy will shake up the world in the twenty-first century."[1]

When asked whether, in his vision of life as a genome sequence, he was missing "life's complexity, the legend in the map," Yang replied, "We are far, far away from knowing life. What we are doing is very superficial and sequencing is only the beginning, comparable to the periodic table in chemistry." Here he indexed the heuristic nature of modern experimental science; historically, the rendering of chemistry in tabular form enabled the standardization of a language between disciplines, which helped to push life science from natural history to an experimental science of life, biology, and, later, genetics.

But the audience was less concerned with a question over the heuristic nature of modern experimental science than the ethical problem of its potential applications. Yang responded to skepticism of his organization by clearly distinguishing BGI's power to rewrite the program of life from the synthetic biology unfolding in the West as exemplified by J. Craig Venter's goal to reengineer life at the cellular level. But perhaps what BGI is doing is not that different from what scientists at Harvard, MIT, and Berkeley are doing, only it is at a more stunning scale.

Yang emphasized that there is a critical East-West difference in the uses and implications of post-genomics biology. In Yang's view, in technologically advanced countries, the public worry is about the harm that science can do, whereas in developing countries such as China, science is viewed as a problem solver, and there is the need to be concerned about the ethics surrounding its uses. Yang seems to be saying that in China, and Asia more generally, the ethical justifications take a different form from "Western" scientific responsibility, which, in his estimation, hinges on the supposition that it is the way scientific applications are used and mobilized that renders whether something is good or bad, ethical or not ethical.[2] By comparison, in the developing country case, Yang suggests, these justifications take a different form. They focus less on the ethics of potential applications and so are not hung up on what appears to be time-wasting committees or scare journalism, but instead are poised toward their usefulness in informing actual and practical decisions, positions, preparations, and coordinated action. In other words, scientists from the People's Republic of China (PRC) are not claiming to reinvent the wheel like the Americans; the claims and aims of BGI, for instance, are not first to do good with science, but to do good science, with the hope of the results of that work ramifying as a collective social good down the line. It seems a more humble ethos in the application of science for the common good. But "good science" in the BGI incarnation is about many things: being a good scientific citizen, and generating good values for human welfare through an efficient system of corporate science.

In this chapter, I discuss BGI in order to illuminate its differences from Biopolis, the main focus of this book. I begin by noting the rapid rise of BGI, a nonprofit institute and company in China, which is not affiliated with the Chinese state but now dominates the world in DNA-sequencing prowess. Unlike Biopolis, which is an extension of Euro-American cosmopolitan science using "Asian" materials, I suggest that BGI is a new model of Chinese global research and business that stirs both skepticism and anxiety among Western observers, both for the sheer magnitude of its sequencing operations and for the ways in which it is taken to embody, rightly or not, a number of worries over Chinese science in its deviation from cosmopolitan scientific structures and strictures. I then shift to the dual faces of BGI, contrasting its international and domestic modalities of biodiversity for research on emerging life forms. On the national front, I argue, BGI deploys ethnic classification in a signal study of Tibetan DNA that seems to foreshadow China's biomedical preparation for global uncertainty.

Not Just a Global DNA Assembly Line

BGI Genomics is a private, nonprofit organization founded as the Beijing Genome Institute in 1999 to participate in documenting 1 percent (the "Chinese" component) of the Human Genome Project. The cofounders, Henry Yang and Wang Jian, along with two other colleagues, were members of the "lost generation" who grew up during the Cultural Revolution (1966–1976). Some of the most entrepreneurial leaders in China today are from this cohort. They went through hardships, including the closing of all schools, but later managed to go to college and even enroll in universities in the West. The future BGI leaders were trained in genomic science: Henry Yang at the University of Copenhagen, and Wang Jian at the University of Washington, Seattle. With their cosmopolitan experiences, Yang and Wang are very unlike China-trained scientists employed in state institutions. BGI started as a nonprofit research entity affiliated with the Chinese Academy of Sciences, but it split off into a private organization in 2007 when state funding dried up. BGI leaders also chafed against the bureaucracy and conservatism of the Chinese state science world that was skeptical of such an expensive science venture. In addition, national agendas did not permit freedom in biological research within the state system. BGI decided to be formally and institutionally independent of the state, I was told by a corporate representative, in order to be free to choose its own projects without risking possible abuses of science.

In 2009, the Shenzhen government offered BGI close to US$13 million to move to its Special Economic Zone, near the border of Hong Kong. Commenting on the move from the nation's capital, a BGI investor said, "Shenzhen is as far from Beijing as you can get. You can't be independent in Beijing."[3] Recast as BGI Genomics Shenzhen, the company operated out of a former shoe factory, enjoying the same cheap land rates and tax breaks as its neighbors, including the giant Foxconn factory that manufactures Apple digital products for world markets. Shenzhen is the technological incubator of China, providing prime opportunities to combine a mass assembling infrastructure with DNA research. As BGI's international sequencing business grew, the company opened BGI Hong Kong in another former shoe factory, putting the company close to global transportation networks that allow a quick turnover in processing DNA samples and performing medical diagnoses for overseas clients. Having grown accustomed to Biopolis's high-tech and resortlike enclave, I was surprised at how basic and unglamorous BGI Hong Kong is, situated in an outdated industrial zone.

Although BGI had divorced itself from PRC political and scientific funding establishments, it has benefited from funds and tax breaks offered by the Shenzhen and Hong Kong governments. By this time, BGI had made the central government proud for putting Chinese genomic science on the world map. In 2010, the Development Bank of China offered BGI a loan of US$1.5 million to purchase sequencing machines from the United States, making it the world's largest facility. The company headquarters at BGI Shenzhen has more than 158 sequencing machines, and it claims to have sequenced some 57,000 human genomes to date.[4] The institute employs four thousand people, including two thousand PhDs. There are over a thousand young employees in bioinformatics alone, many of whom live in company dormitories. But BGI Shenzhen has grown beyond being a sequencing platform to designing new medicines and food products. Institutionally, BGI Shenzhen is a bit like the Biopolis campus in that it has different divisions, dedicated to diagnostics, animal cloning, and agricultural research.

In a bold innovative move, BGI in 2013 acquired the assaying company Complete Genomics in Mountain View, California, and now has the capacity to produce the machines that produce the data. It is estimated that BGI has at least 25 percent of the world's total gene-sequencing services, followed by Illumina of San Diego and the Broad Institute of Harvard and MIT.[5] In the world, BGI is mostly known for being the world's largest sequencer of genetic data on animals, plants, microbes, and humans, giving BGI the capacity to shape the evolving global ecosystem of genomic science.

The meteoric rise of BGI has stirred trepidation in the world of bioscience. For observers in the West, BGI has been viewed as the apparent spitting image of the PRC industrial behemoth, literally built into former factory spaces hiding in plain sight in China's industrial zones. The uncertainty for those on the outside is that BGI is a chimeric entity; this view is heightened because, with China's ascending global economic power, "security" concerns are the expression of a suspicion over the categorical hybridity of Chinese institutions in general, and of the gigantic capacity to dominate global industries from manufacturing shoes to manipulating genomes. The narrative of Singapore's Biopolis—U.S.-influenced, cosmopolitan, capitalist—contrasts with BGI's image as one of those strange PRC hybrid entities, a chimera of socialism with Chinese characteristics: Is it or is it not a factory? Is it a private or a state entity? Is it a research or a capitalist institution? As a biomedical milieu, is it doing ethical or unethical things?

In foreign science journalism, the digital mechanization that drives Shenzhen's industrial powerhouse has been transposed to the fast informatization of living forms. In 2010, a writer for *Nature* magazine dubbed BGI "the sequence factory" and skeptically asked whether "its science will survive the industrial ramp-up."[6] After BGI's purchase of Complete Genomics, a *New Yorker* piece repeats the factory theme, branding BGI "the gene factory."[7] An image of "assembly-line" DNA is used to describe BGI's global reach.[8] From Western journalistic and science perspectives, calling BGI a genome factory implies that because of its Foxconn-like, mass-assembling approach to data, there is a skepticism as to whether the company can be an innovator in science.

Dr. Svensen, a geneticist, previously at Biopolis but now at the University of California, San Francisco, remarked to me, "Genome sequencing is just a global service, that is, stupid work that should be industrialized. Once sequenced, it is up to the scientist to analyze how genomic information is different." His comments suggest that China's science power lies in its cheap labor, not intellectual creativity; there is also the suggestion in factory imagery that workers are exploited in assembly-line data production. At the same time, Svensen seems to miss a different truth, which is that genomic sequencing is a platform for scientific experimentation. While the genomic sequencing infrastructure is not the current aim of the science, a monopoly of global data points exerts a kind of biostatistical power to monopolize markets, to write new algorithms, to plan a novel design of life.

There is trepidation about whether BGI is a state-driven institution that challenges how international science is conducted. Related to this worry is BGI's fusion of research and business in a new kind of global science facility. When I asked Dr. Chen of BGI Hong Kong to address such criticisms, he said emphatically that BGI is not a state agency but "a nonprofit and a commercial venture, a research and a marketing project." He went on to say that BGI may be "hybrid" but not in the sense of being a joint state venture; instead it is like any American private company (like Google) that vertically integrates

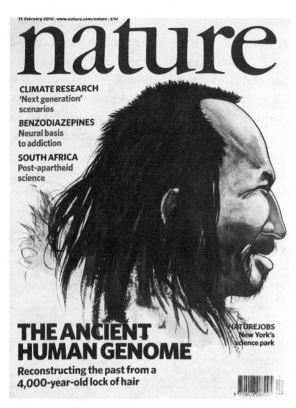

FIG. 9.1 BGI's study of the "ancient human genome" makes a splash. Courtesy of *Nature* Magazine.

multiple research and business units. As such, as the world's largest genomics center, BGI is attractive to investors worldwide, including the Silicon Valley venture capital firm Sequoia Capital. At Berkeley, Dr. Rasmus Nielsen, an evolutionary biologist who works closely with BGI (see below), noted that anxiety about BGI as a "genome factory" should not be about BGI as doing something unethical. Rather, there has been concern about the speed of BGI's rise and its ever more complex logistics and bioinformatics, all factors that decisively inform its global competitiveness. But, as we shall see, BGI is a new kind of science company that is innovative on different fronts.

The "sequence factory" label was first earned when BGI analyzed the DNA of an ancient human from a hair fragment found in Canada's ice wastelands. Featured on the cover of *Nature* magazine (February 2010), the study put Chinese life sciences for the first time on the global map (see figure 9.1). The arresting portrayal of an ancient human became a kind of ethical branding of BGI. In its plan to digitize, eventually, the entire human pool, proceeding from Asia to Africa and South America, BGI has won another moniker: "a library of digital life."

As an ambitious science organization, BGI has been innovative in forging international collaborations with major research institutions and joint labs in the West. Projects with the University of Copenhagen, where Yang trained, include the sequencing of the Danish pig and the study obesity in Denmark. Under the umbrella of BGI Americas, in Cambridge, Massachusetts, BGI is helping to build a DNA analysis center at the Children's Hospital of Philadelphia, and it is developing programs in food security and human, animal, and environmental health at the University of California, Davis. There are new BGI branches in South America and Africa. As a global sequencing powerhouse, BGI is a critical provider of bulk services to U.S. institutions.

BGI has also deployed its sequencing capacities in its role as a global citizen of science. After the Asian tsunami in 2008, BGI experts sequenced the DNA of victims to help with the identification of their nationalities (aligned with ethnic profiles of their DNA). In 2011, after a mysterious outbreak of food-borne diseases in Germany, BGI sequenced the *E. coli* strain found in contaminated sprouts within three days and made the data freely available, which helped to put an end to the contamination. When it comes to Asia, BGI offers scientists, including those at Biopolis, reduced costs for sequencing services as a way to boost their research.

Indeed, genome sequencing on a large scale is a relatively easy way to achieve a global presence for Chinese science. As a commercial enterprise, BGI has been an inexpensive and speedy sequencer for researchers around the world. It is a very complex, multifaceted genomic science enterprise, with different divisions focused on the technologies of sequencing, screening human DNA for medical applications, and developing plant and animal hybrids for food security. BGI promises to change the infrastructure, business, and innovations of cosmopolitan science. In an email, a BGI manager describes the following scenario: "China is rapidly positioning itself to become an important—and hugely disruptive—player in the industry's future trajectory."[9] By putting its sequencing prowess at the service of the world, BGI has already made an impact in sequencing the planet's biodiversity.

Modalities of Biodiversity

A question for an imperiled world today is whether we should value the whole of biodiversity for its own sake or for the differences composing that diversity. DNA sequencing is therefore about mapping life on earth and, in the process, discovering findings that can sustain the health of our species and the planet. I suggest in the rest of the chapter that BGI, in building its sequencing databases, prioritizes two models of biodiversity by using different metrics of species and ecological levels. I compare BGI's modeling of the "tree of life" in two ways. The first universal tree prioritizes species that are important to economy and science (i.e., values of the ecological sustainability of our planet), while the second "Chinese" tree identifies the scientific and aesthetic values of iconic species within the cultural-ecological habitat of the national motherland.[10]

BGI's approach to the book of nature is to model DNA databases in terms of their specific scientific findings, but it makes a distinction between the general planetary biosphere and a Chinese biosphere. The classical image of the tree of life is up for revision in BGI's vision. At the conference where Yang encountered skepticism from the audience, he professed his company's goal of "flying the science and humanity banner," an intention he expressed in a flamboyant style.[11] In 2010, BGI launched the 1000 Genomes Project in order to generate reference genomes for a thousand "economically and scientifically important plant/animal species."[12] On the webpage, a tree of life (see figure 9.2) diagrams a certain logic of assemblage in that *Homo sapiens* is not at the top of the tree but ironically reoccupying a position near the center of the biospheric tree. The trunk of this global tree is a double helix, suggesting that the species leaves are related because of the various combinations of nucleotides in DNA rather than the splitting branches and bifurcations of evolution. Indeed, there are no branches at all, suggesting a biosphere or atmosphere of life rather than the arborescent tree. Also there are openings to the sphere, which suggests more worlds of life and form beyond the selected one thousand genomes of flora and fauna important to human beings. What kind of anchor is the modern human positioned therein, relative to the surrounding circles of animal and plant species spinning on the top of a blue planet Earth? As a figure of both immanence and transcendence, this tree of life is haunted by the intertwining interests of ecological sustainability and corporate branding. The diagram depicts the interrelationships among different animal and plant species, suggesting that genomic findings would yield tools for sustainability that can be economically made accessible to all of humankind.

FIG. 9.2 BGI's tree of life for the 1000 Genomes Project.
Courtesy of BGI Shenzhen.

Meanwhile, within the lobby of the BGI office in Hong Kong, there is another tree of life (see figure 9.3), one with the particular Chinese lens of biodiversity within China's particular ecological, cultural, and political sphere. The intrusion of a Chinese biocosmology into DNA mapping is perhaps unsurprising. BGI scientists, especially the president Wang Jian,[13] present themselves as patriotic citizens of the PRC who want to do science that contributes to China's sustainability, prestige, and national identity. There is a race to sequence the DNA of humans, animals, plants, and microbes, that is, to mobilize the knowledge of life forms considered part of the national patrimony. The project is not driven by the state, and indeed Chinese state science institutions are neither coordinated nor entrepreneurial in the way that BGI is.

But beyond its resolutely international orientation, BGI has a homegrown interest in building a Chinese genomic treasure house that can contribute to

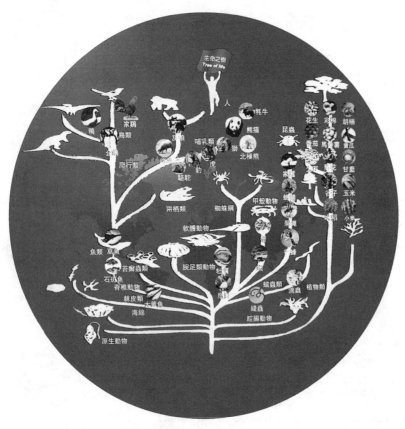

FIG. 9.3 BGI's other tree of life. Photograph by Rena Lam; courtesy of BGI Shenzhen.

the health and biosecurity of the nation. Thus, while aiming to model the entirety of life, BGI also focuses on generating genomic maps of Chinese forms of life, especially charismatic species like the giant panda, the stork, and the silkworm, as well as flora (soybeans, golden ancient poplars, mushrooms) that are specific to the Chinese ecosystem, which is coextensive with China's modern territorial boundaries. Nonnative animals and plants—chicken, rice, peanuts, tomatoes, maize—have long been Sinicized as food crops vital to the civilization.[14] The China-centered ecosystem emphasizes differences between life forms in order to generate commodifiable value for research and medicine. The increasing facility and speed of the bioinformatics software have greatly improved Chinese scientists' capacity to respond to major arenas of concern for the PRC: to develop genetically modified foods (e.g., rice and maize) and cloned livestock and to develop stem cell research to treat human diseases. The first step in wielding this bioscience prowess is to stake scientific and symbolic claims on China's charismatic and necessary biological forms.

The BGI's Chinese tree of life, a projection of the artist's imagery, thus enacts China as a distinctive genomic branch of the tree of life. This branch image suggests a very different conception of evolution and relatedness among species than the amorphous DNA cloud depicted in the other tree model. The unified origin of Chinese species, where people (at the top) branch off from primates, first suggests a unified descent in place, which seems to suggest a deep territorial-evolutionary relationship to China as a historical-cultural complex. Besides native species, the tree incorporates nonindigenous ones that are historically part of the glorious Chinese food culture. The Chinese tree is very different as well from the more conventional Biopolis notion of populations resulting from the migration of an already-evolved human species.

This powerful representation of Chinese life forms evokes not only their innate qualities but also their cultural and even mythical roots and routes to China's present, with iconic species mapped onto the flat time of national culture heritage, ending with a Chinese-identified human figure at the top. Mimicking the aesthetics of classic porcelain design, this diagram contains and delimits Chinese forms of life as so many branches of a nationalized multispecies tree within the shape of a blue plate. As a patterning of genomic truth claims, the tree metaphor embodies the very singular *oikos* of an emergent globalizing Chinese ethnos.

Ethnic Classification and Governing through Blood

As mentioned above, BGI has bigger ambitions than PRC state-run science institutes, but BGI also has its own dream for Chinese biomedical science. It has turned its algorithmic power toward differences composing human biological diversity in China. The goal is to develop medicine that can be customized for different groups in the country.

There are, however, significant differences between BGI and Biopolis, in the deployment of the ethnic heuristic and itineraries of medical information. Elsewhere, I illuminate how researchers in insular Singapore use the ethnic heuristic for constructing DNA databases that can create broader environments for making the categories fluid and fungible. The ethnic-specified digital knowledge makes ethnicized DNA convertible and substitutable across different domains of science valuation so that these objects can represent majority "Asian" populations in the world.

BGI's participation in racialized medicine has different international and domestic aspects. Internationally, it provides information for the Asian Cancer Research Group, a nonprofit company that investigates cancers that are prevalent in Asia, primarily focusing on liver, gastric, and lung diseases. Eli Lilly in Singapore, with Merck and Pfizer, takes the lead in organizing the collection of profiled tumor samples and data throughout Asia. Big pharma is clearly aligned with racialized medicine. A Pfizer scientific officer explains that "environmental and genetic factors are believed to underlie the dramatic differences in the molecular subtypes and incidence of cancers in Asia and other parts of the world. Although some progress has been achieved in the last few years in understanding and treating these cancers, they remain a huge unmet need and a disproportionate health burden to Asian patients."[15] BGI's is the first-of-its-kind genomewide study of recurrent hepatitis B virus that causes the most common form of liver cancer in China, with the highest rates in the world. The Asian cancer group is a new trend in which big drug companies engage in a precompetitive collaboration, combining their resources and expertise to accelerate research of disease and disease processes.

While the Asian Cancer Research Group uses the "Asia" marker as an immutable mobile in a Biopolis-like manner, BGI's own projects on genetic diversity and ethnic differences are about identity in and of place, as classified within the territory of China. Whereas Biopolis projects deploy ethnic-differentiated DNA in an expansive, origami-like digital configuration of "Asia," BGI's projects identify ethnic DNA differences as points of encounter in the stream of flows all firmly bounded by China's official history and borders. The comparative ethnic DNA mapping in China is about the social ordering of ethnic differences and associations that are not substitutable outside the Chinese world. Below, I will discuss this China-centric orientation in the Yan-huang and Tibetan projects. But, first, a brief account of China's pervasive ethnic classificatory scheme is necessary.

Human sciences in the PRC are institutionally obliged to follow the system of official nationalities. Since the sequencing of the human genome occurred, PRC-born researchers, following the official ethnic classification of the Chinese nation, have used Han Chinese as the master ethnicity. The Ethnic Classification Project (*minzu shibie*) of 1954 determined the fifty-six ethnic nationalities (*minzu*) entitled to political representation within the territorial expanse of the PRC. These fifty-six *minzu* compose a single master nationality, "Chinese" (*zhonghua minzu*), making China a multinational nation, of which people identified as Han make up the vast demographic majority. Thomas

Mullaney argues that a pre-1947 British imperial sociolinguistic taxonomy for classifying groups in Yunnan, and Stalinist criteria for the categorization of "plausible communities" becoming nationalities influenced the PRC classification of *minzu*. By 1984, a definitive and nonmodifiable fifty-six *minzu* classification was completed, thus establishing a primordial model. Official discourses entrenched this *minzu* scheme as central in the maintenance of the territorial, political, and economic integrity of the country.[16] This scheme highlights two important details: the differences between majority Han (*hanzu*) versus non-Han populations, and the "official-national identity" (*zhonghua minzu*) that is the umbrella of the multinational state. Not surprisingly, scientists are institutionally bounded to work within this official fifty-six *minzu* framework in order to sample and conduct DNA research.

The conflation of the official *minzu* classification and microevolutionary theory seems officially fortuitous in China. Human evolutionists identify an epigenetic rule of gene–culture coevolution that correlates groups evolving in relative isolation with a susceptibility for genes for certain diseases. One may perhaps trace the birth of Chinese genomics to a 1994 project initiated by the Chinese Academy of Medical Sciences to assemble the "immortalized cell lines" of different Chinese populations. A group of geneticists and ethnologists from leading universities (at Beijing, Harbin, Kunming, etc.) set out to collect "relatively pure genes" of "isolated" minority groups on the continent. The Chinese scientists feared that such gene pools were increasingly diluted through exposure to other populations. Dubbed "the world's largest ethnic DNA bank," this state project provides a kind of baseline for subsequent ethnic-associated genome studies in the country.[17] While state institutions individually pursue such DNA research trends, there appears to be no unified state coordination simply because the China science milieu is profoundly under-regulated. Rather, researchers have been impelled by sociopolitical beliefs in ethnic differences and patriotic zeal to pursue convergent projects that give shape to an emerging racial biomedical science.

BGI dipped its toes into ethnic-specified medicine when it joined an international effort to establish the most detailed catalog of human genetic variation ever assembled. The international 1000 Genomes Project, launched in 2008, includes BGI Shenzhen, the Wellcome Trust Sanger Institute, Cambridge University, and the National Institutes of Health in Bethesda, Maryland. The consortium aims to sequence the genomes of at least one thousand anonymous participants from different ethnic groups. In the process of making a detailed map of human genetic variation, the goal is to find rare genetic variants related to diseases. With BGI taking the lead, China became the first country to begin to sequence the whole genomes of larger numbers of individuals. At that time, worldwide, only two individuals had had their genomes sequenced: James Watson and J. Craig Venter. BGI has since sequenced the genomes of two Chinese individuals, one of whom paid about US$1.4 million for the analysis. By 2014 BGI, with its accelerating sequencing powers, had exceeded the one thousand genomes limit and hoped to expand to a one million genomes project for human beings, as well as for animals and plants.

For the past decade, (non-BGI) Chinese geneticists in Chinese universities have been busy analyzing the DNA of the Han nationality, which is genomically distinct from related ethnic minorities in Southern China. A study by PRC scientists calls the Han Chinese "the largest single ethnic group in the world, consisting of ten branches." One study of the Y chromosome and mitochondrial DNA demonstrated, the researchers claim, "a coherent genetic structure of all Han Chinese." Researchers identify an "older branch of the Han Chinese" in the Pinghua group that is represented by ethnic minorities (*shaoshu minzu*) in Guangxi Province (the Dai, Hmong-Mien, Zhuang, Kam, Mulam, Laka).[18] The majority of these scientists are not linked to BGI, but the research institute has begun in a more targeted way to map the genomes of Han and non-Han populations. For instance, BGI has sequenced the genome of a Mongolian subject, said to be a thirty-fourth-generation descendant of Genghis Khan (Mongolian is one of China's official nationalities).

In the international one thousand genomes effort, BGI's contribution is called the Yanhuang Project, which has sequenced the entire genomes of one hundred Han Chinese individuals. The Yanhuang (YH) genome map shows the relationship between YH genotypes and phenotypes and their associations with fatal diseases that threaten ethnic Chinese populations. By making genetic maps of populations in China, BGI provides an unprecedented biomedical resource for developing personal medicine that promises to benefit ethnic Han Chinese. The BGI Shenzhen website posted the following claim: "We Chinese people have our own genetic background, disease susceptibilities and drug response, which differ dramatically with other populations. For instance, Caucasians are reported to suffer more from skin cancer while Chinese suffer more from liver cancer."[19]

The mapping of DNA along the *minzu* axis cannot help but more firmly intertwine the search for ethnic-specified health vulnerabilities with the politics of Han-majority rule. Leading China anthropologists have argued that the post-1949 invention of ethnic minorities not only distorted the past but also involved

Han civilizational attempts to impose dominant values while heightening the sense of ethnic differences.[20] Louisa Schein has explored the postmarket reform's "minority rules"—expressed in media, identity performances, and tourism—as an ongoing cultural production of "internal orientalism," especially with regards to the Miao as a "feminine other."[21] Nevertheless, cultural and linguistic differences between Han and minority groups, especially the Tibetans, may be more blurred than official representations would have us believe.[22] Besides Tibetans, powerful disenfranchised minorities such as Muslim Chinese are constructing their own "ethnic nationalism," in a protracted politics of center-periphery struggles through which the Han majority come to define themselves.[23]

Not surprisingly, an interethnic genomic database of China is both contextual and performative, in that it remediates *minzu* as a biological form and interrelationship, even as "race" was only one of several criteria in the original official classification of groups, tied as it was to conceptions of ethnicity adapted from Stalinism and its conceptions of social-political evolution rather than biological unity. Ethnic-differentiated genetic data perform knowledge affects, drawing on the authority and social order within which they are produced. After all, the Yanhuang name is historically and culturally extremely significant, for it is the conjoined names of mythic ancestors of the northern (Huang) and southern (Yan) branches of an ancient group, the Huaxia, who are believed to be ancestors for the Han peoples. In this chauvinist move, substantive genetic unity is created out of the national origins story that traces Han roots in historical groups from the northern and southern halves of this vast continent. Thus, BGI's one hundred genomes project for Han Chinese and other studies genetically establish the Han Chinese as the original national population in a genomic majority-minority scheme of ethnic groups.

When I discussed the Yanhuang name, the historian Wang Gungwu said that trouble begins when geneticists use historical names out of the fog of history to designate the historical and genetic originary compositions for contemporary groups in China. This sort of biologization of a deep cultural-historical memory is a technological aspect of what Benedict Anderson calls "imagined communities," a prerequisite and ongoing process in the creation of nationalisms.[24] Wen-ching Sung has argued that the "imagined national-ethnical identities turn genomic research into vehicles for recapitulating and substantiating the notion of Chinese ethnicity."[25] Han ethnicity (and the broader ethnic scheme of the PRC) gains substantiation through bioinformatics, and this not only shores up hegemonic racial formations but also aims to establish the long historical-national claim to an antique Han domination and racial unity. The YH project propagates a genetic consciousness that reinforces beliefs in the biological "sameness" of diverse ethnic communities gathered as "Han" in China. The YH project, Dr. Chen of BGI claimed, is only a "primary model" in building "the China Genome data bank as a national bank." The main concern is to find disease susceptibility genes and learn more about hereditary diseases as a way to maintain the health standards of Chinese people. It seemed reasonable to focus first on the Han as the largest ethnic political entity, Chen continued, but other groups will be included in future DNA studies. Ethnic-differentiated genetics is not merely ideological but constitutive of a new way for managing the biological health of the nation.

Besides the multi-*minzu* political order, the rise of genomic science in China also requires a new mode for accessing human samples. After generations of blood-donation public health campaigns, Vincanne Adams, Kathleen Erwin, and Phuoc V. Le have argued, contemporary China has increased its supply of safe, transfusable blood. By compensating blood donation that is not commercialized, "reciprocal obligations between citizens and the state are managed in and through blood." A systematic program of "governing through blood" ruptured traditional notions of blood as a precious family essence or *qi* (spirit or energy), and it infused a popular embrace of blood donation as "an act of altruistic patriotism."[26] Organized by workplace units (*danwei*), obligatory compensated blood donations (starkly contrasted to "blood selling") have become a normalized way to participate in socialist welfarism. With biology politically actualized by this mode of blood governance, work-unit quotas for patriotic blood donations would elicit samples from *minzu* groups across the nation.

This mode of Chinese biopolitical governance thus opens channels to blood samples, allowing genomic scientists, including those in BGI, to create ethnic-specific genetic mappings through which people can be analyzed and thus administered for the well-being of people in China. By banking Chinese genomes, Chen explained, BGI is building the foundation of "a prevention model versus a disease model." The reasoning seems to express a kind of preemptive social eugenics propelled and necessitated by China's demographic heft and anticipated health, social, and political problems in the near future. Chen indicated that this preventive approach was especially urgent because China's one-child policy has greatly increased the burden—the psychological, social, and individual costs—that the younger generation bears in caring for their aging parents. Large-scale genetic studies of China's populations, he reasoned, would yield potential DNA information that can inform medical solutions to anticipated genetic health problems and thus ameliorate associated social effects among second and future generations.

At the same time, this preemptive strategy makes use of a cross-ethnic DNA comparison to track the differential distribution of biological weaknesses and capacities across ethnic groups scattered across the vast continent. The operating logic is ethically problematic, rooted in the supremacist pragmatism that any genetic weaknesses identified in the Han majority can be potentially rectified by analyzing genetically beneficial traits found in ethnic minorities. To this end, BGI has multiple branches throughout China, an institutional distribution that mirrors the territorial location of significant minority groups and research topics: Hangzhou (livestock, plant, and health genetics), Xishuangbanna (Dai and other minorities, tropical biodiversity), and Lhasa (Tibetan). At the broadest level, BGI is combining research on biodiversity in plants, in animals, and of human beings almost as iconic species of a singularly China-specific biosphere.

Tibetans and Peak Performance

In the BGI contribution to the one thousand genomes project, the institute first did the Yanhuang study of Han DNA and then moved on to the sampling of Tibetan DNA. The decades-long history of Tibetans' struggle for autonomous rule from the PRC has made them a politically potent people at home and abroad. But BGI's framing of the comparative genetic study seems almost whimsical, not political. Jian Wang, the charismatic president of the institute, had picked up mountaineering while he was a research fellow at Seattle University in Washington (he did postdoctoral work at the University of Texas and the University of Iowa). Returning to China, he became a serious mountain climber and has successfully scaled Mt. Everest. On his climbs, he developed a personal and professional interest in the different capabilities of Han and Tibetan hikers. He was reported as confessing, "I have found that Tibetans are much better than all of us [Han Chinese] on the high mountain, and I wanted to know why."[27]

This desire materialized in a BGI Tibet-Han project—comparing allele frequencies correlated with adaptations to high altitudes—that was published in 2010. BGI researchers identified fifty Tibetan villagers living above an elevation of 14,000 feet (where there is 40 percent less oxygen than at sea level) and gathered blood samples in order to analyze oxygen saturation, red blood concentration, and hemoglobin levels (see figure 9.4). The peak performance

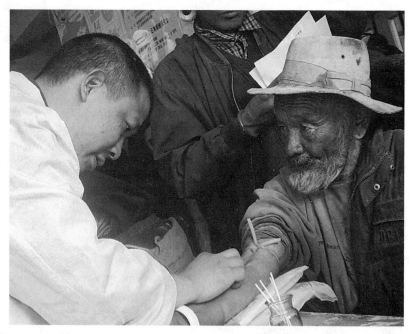

FIG. 9.4 Collecting a blood sample from a Tibetan subject. Courtesy of BGI Shenzhen.

of Tibetans in the Himalayas then was compared with the lung capacities of forty lowland Han Chinese subjects from Beijing.

The Tibetan-Han data then shifted to Rasmus Nielsen, a professor at my home campus in Berkeley and a collaborator on the project.[28] His subsequent computational findings using BGI data received a lot of media attention for their evidence on human evolution. Nielsen and his team analyzed the genes of fifty Tibetan individuals and identified thirty genes with DNA mutations, including a mutation for the EPAS1 gene, which is linked to lower levels of hemoglobin. The EPAS1 gene seems to restrain the overproduction of red blood cells at extreme altitudes, endowing Tibetans with greater resistance to altitude sickness than other groups. The EPAS1 mutation and physiological mechanisms for high-altitude hypoxic adaptation were much less prevalent in the DNA data derived from the Beijing Han sample.

Nielsen shared his view that most geographic variants in anatomy and physiology (often culturally identified as "racial") are due to "genetic drift" or random fluctuations in gene frequency, or the migration of mutation-bearing individuals to other sites. Populations in highlands are good gene pools due to their geographic isolation, but they do share genes back and forth with other groups. By comparing the SNP (single nucleotide polymorphism) analyses of the Tibetan and Han samples, he determined that the two groups, which share many genetic traits, diverged nearly three thousand years ago. For Nielsen and his team, as well as BGI, the Tibetan study is a coup, and it established their findings as indicating the fastest case of environmentally driven genetic micro-evolution in a human population. In a 2014 report, Nielsen and colleagues traced the ancestry of the EPAS1 variant to relatives of early humans, the Denisovans (contemporaries of the Neanderthals) in Siberia.[29] The Tibetan DNA study is a triumph for the population geneticists, and it was well featured in the American media as a stunning case of natural selection and human evolutionary adaptation.

But given the politically charged nature of Tibetan-Han relationships, and the political implications of genetic Han-Tibetan comparisons, this study is not permitted to be solely about phenotypic plasticity. Therefore, despite the scientific celebration of their evolutionary prowess, Tibet scholars and leaders have rejected the genetic findings, especially any claim of Tibetans' descent from Han Chinese. Critics pointed to evidence that the culturally identified Tibetans have lived on the Tibetan plateau for more than ten thousand years, far exceeding the timeline of divergent population evolution offered by genomic researchers. Robert Barnett, a scholar of Tibet at Columbia University, was quoted as saying that Tibetans viewed the findings as a strategy to provide scientific evidence that Tibet and Tibetans were integral parts of Han China as a race, a people, and a nation. Seeking to diffuse such fears, Professor Nielsen defended himself in the press: "What identifies a people isn't genetics, it's cultural heritage. I don't think this study has any implications for the debate about Tibetan independence and their right to self-determination."[30] From the perspective of BGI scientists, the significance of the findings was not sociopolitical but pharmaceutical, which, of course, has its biopolitical weight as well.

In his campus office, Nielsen remarked that, as an evolutionary biologist, his main interest lies in figuring out genetic exchanges between continents by tracking the DNA of early human migrations to China. Therefore, to him, "the Tibetan case regarding adaptation is great. . . . Not all have direct applications for evolutionary biology. It's a bit like studying different kinds of birds in migration." That sounded like a faux pas, comparing Tibetans to birds, but Nielsen was well aware of anthropological unease over the conflation of cultural and genetic collectivities, and of political charges of potentiality of eugenics. He expressed frustrations at damage done by American media that sometimes misrepresent bioscientists as thinking that "race is real," as a *New York Times* journalist had said. This suggestion that scientists traffic in biological essentialism and are unaware of potentialities of eugenics research is "nonsense," at least for Euro-American experts. After all, Nielsen pointed out, Euro-American science culture is permeated by an anti-eugenics ethos, "but debate has not yet begun in BGI."

The EPAS1 gene is an evolutionary adaptation to compensate for decreased oxygen level, or an example of phenotypic plasticity, not genetic change. In this study of evolutionary adaptation, Nielsen remarked, the critical difference lies in geography, as in his analysis where Tibetan genetic adaptation to high altitudes was measurable in different hemoglobin responses to thin air. Instead of using ethnic categories, he could have as easily substituted "highland group" for Tibetans and "lowland group" for Han Chinese in the two data sets, replacing a state ethnic heuristic with an ecosystem one. But BGI researchers had already conducted their sampling by using self-identified ethnic groups. For now, ethnic designations are a convenient gloss for population differences, one with implicit state endorsement and through which BGI can explicitly align its research with the well-being of extant groups in the tapestry of Chinese ethnicities.

Nielsen suggested that in the future geneticists will move beyond using ethnic collectivities because within each group "the genetics may not be the same." When I asked him whether ethnicity has been deployed in genomic research as a shortcut to DNA variation, he demurred. Now that large-scale genomic data have been compiled, he added, "We can throw away the ethnic association and go directly to the genetic variant, the high rate of frequency associated with it." He stated, with some heat, that an "ethnic differentiated database is *not* that useful . . . it is a choice." The statement is sufficiently ambiguous so that one may surmise that BGI ethnic-specified DNA information is still somewhat useful because, after all, he went along with the project, even though only at the point of computational analysis and design. Is this an American way of demurral, a vulgarization of science to the public?

Perhaps Nielsen is being disingenuous here, refusing to associate with the ethicized mode of sampling while enjoying the prestige and other benefits that came with working on such a trailblazing project. After all, the pathbreaking nature of the Tibetan project depends precisely on the determination, genomically, of the moment of ethnogenesis, so that it would need ethnic groups (Tibetan, Han) to make comparative sense. When ethnicities are determined by specific gene variations, can one continue to speak in terms of "populations" in disassociation from the official ethnic labels? As a BGI collaborator, Nielsen seemed to participate in the use of ethnic identifiers as a way of complying with Chinese political necessities in order to gain access to the samples. It would be challenging indeed to disentangle himself from the official Chinese ethnic matrix that provides the conditions of possibility for this experiment.

Air Hunger

Outside the lab, the EPAS1 variant is called the "athlete gene" for its link to a physiological trait—the increased production of hemoglobin—that is key to physical performance in high altitudes. Therefore, the Tibetan DNA project may be considered a strategy of the BGI ambitions for pharmacogenomics of practicality and efficiency, values one associates with an earlier era of confident, high modern Western science. But back at BGI, Chen framed the comparative ethnic DNA study in terms of the science governance of the nation's peoples. Unlike his colleagues educated at leading Western universities, Chen received his medical degree from a university in Hong Kong. He worked as a businessman selling British and U.S. biomedical equipment in China before joining BGI to develop the Hong Kong site as its international business hub. Chen may be more prone to blunt statements, speaking without the nuances that one finds in the remarks of Wang Jian and Henry Yang, the more cosmopolitan and culturally adroit leaders of BGI.

Invoking natural selection and genetic adaptation, Chen said that "populations in highlands have good Asian gene pools due to their geographical isolation. In isolated territory, we can consider the gene pool [to be] more conservative than other Asian genes; the idea is that isolated populations hold onto their genes better than us" (i.e., the Han Chinese). Because isolated populations yield genetic diversity, BGI conducts DNA analysis among them to "prove that the environment will create survival benefits . . . to have stronger confidence to say that in that environment, genes rise to the challenge, or select for it." He pointed to the finding that Tibetan genetic evolution has tended toward the release of more oxygen in their oxygen-scarce, high-altitude environment. BGI researchers hope to mimic the hemoglobin-bonding oxygen molecule discovered in Tibetans so that they can develop therapies for people without the genetic and physiological adaptations for living in high altitudes.

I expected to hear about the anticipation of profits stemming from the discovery of the athlete gene. So Chen's response caught me by surprise, for its metric shifting from mere biological enhancement to capacitation of populations. Because of "climate change," Chen explained, "there will be depleted oxygen in the future, a hundred years from now." There is a need to develop this novel medicine based on Tibetan DNA mechanisms "so that other people without this mutation will not suffer from pneumonia, headaches, and so on," if they should be forced to move to high altitudes or when the air around them thins. Suddenly a whole new vista comes into view: the need of biosciences to address China's looming uncertainties.

Here was a vision of genomic science as oriented toward anticipated eco-logical catastrophe. The focus is on non-Tibetans, who may need to move to the Himalayas or cope with living in climate change–induced, Himalaya-like air conditions. The projection is that biomedicine will help the Han Chinese majority, who may migrate to the highlands of their country in greater num-bers, as many are already doing, with more physiological capacities than they have in their lungs to cope with poor oxygen. In this framing, the Tibetan study is a simulation of a climate-driven future in which the anticipated bio-threat is not infectious diseases but uninhabitable lowlands.

In this scenario, BGI's comparative ethnic DNA approach suggests that differ-ent groups hold onto and conserve different kinds of genetic benefits for coping with a precarious future. The reasoning is that ecological isolation of the Tibet-ans has kept hypoxia adaptation from undergoing genomic dilution, optimizing them for their current mountainous zone. This biological advantage can pro-vide clues to medically help other groups who are not so endowed. The casting of Tibetan's hypoxia as a kind of optimized extant, where Tibetans represent a better-prepared genetic-physiological "type" of Han, a Han future-body, has profound biopolitical implications. Han peoples come to depend on Tibetan genomes—not as distant ancestors yoked by genetics to a primordial nation that has been Chinese all along—but as a genomic resource for a coming en-vironment. That the athlete gene has an extraordinarily specified potentiality in solving China's demographic and health problems in a climate-transformed future is itself rather breathtaking.

The PRC Genomic Analog

The sheer sample size of the genomics data that BGI is able to collect (with po-tential collaboration with hospitals throughout China) is the foundation to the company's global power as a research engine. BGI core researchers control and own the patents to their findings, but they also work collaboratively as a way to obtain intelligence, while still maintaining controls over the science data. In sum, the massive sequencing power of BGI, as well as its growing monopoly of DNA data on plant, animal, and human populations worldwide makes it hard to ignore as a global scientific presence. With high-throughput sequencing tech-nologies, BGI has laid an ambitious bioinformatics infrastructure that changes medicine from hypothesis-driven to data-driven. Such large-scale mathematical modeling is considered a necessary advance for multilevel research at the ge-nomic, the epigenomic, and the molecular scales. These are steps toward the de-velopment of customized, cell-based medicine, pursued through international partnerships to study autism, obesity, cancers, infectious diseases, and also brain disorders. In other words, the mathematical model for analyzing gene behavior has become strategic for developing molecular interventions for treating he-reditary diseases and shortcomings. Perhaps not surprisingly, the corporation faces ongoing skepticism as to its quasi-industrial approach to bioscience, the quality of its science, the role of the state, and its politico-ethical goals.

First, BGI leaders such as Henry Yang are well aware that the corporation operates under a cloud of international fear of China as a rising science power. When I interviewed him, Chen candidly admitted, "As it is, the Chineseness of BGI already raises suspicions; sometimes people think that we are a PRC state agency. There is also skepticism that maybe we are not so smart or good at our work." Therefore, BGI scientists insist that BGI is more than a factory, an assertion that perhaps echoes a growing dissatisfaction in China's current development model with being the world's workshop (home to outsourced, labor-intensive, grunt work) rather than being a bona fide center of innovation, scientific or otherwise. Western perceptions of BGI's "factory"-ness, and its as-sociated image of being entirely profit-driven, are a mode of anxious dismissal of China's increasing scientific capacity and place in the global landscape of biological research. There is also Western anxiety over "science" coming out of China as being allied, always potentially, with the state and thus always po-tentially tainted. Statements of concern about BGI seem to be about fitting China into a suspicious slot, where the work is suspect, despite the company's demonstrated competences in corralling data on multiple life forms.

Second, there are misgivings that such a huge Chinese biotech corpora-tion may be engaged in redesigning life itself, fueled perhaps by Hollywood scenes of fiendish Chinese scientists (e.g., Hollywood's depiction of *Dr. No*) taking over the world. I therefore asked Chen to compare BGI to the J. Craig Venter Institute, another major private bioscience company based in Califor-nia. Chen said that BGI projects are "more natural, that is, focused on practical things. We want to avoid projects with uncertain outcomes and that will raise global controversies." By "natural," the focus is on what makes political and scientific sense to improve life and living. There is also a vision of the future and of life and (natural) sciences in accord in China. The overriding ambition of BGI, he stressed, is to put all life on earth on the digital map (it currently produces a quarter of all genomic data), which BGI promises will be made freely available worldwide. Chen reiterated the "humanitarian" goal: "We are interested in things that can bring direct benefits to mankind: issues of illness, health, preventive medicine, even helping victims of natural disasters." I left BGI with the sense that, despite their good intentions, scientists there have not considered the question of whether bioinformatics is, at its core, an emerging enactment of life rather than merely a reduction of life to information.

Third, Chen said, "Bioinformatics assemble the unknown," compared to a "known." I was reminded of Slavoj Žižek's warning about "unknown knowns," or things we do not know that we know.[31] I therefore asked Chen how researchers at BGI decide to accumulate the unknowns, which they already seem to know in advance. "Sometimes," he responded, "disease is the prompt, its spread among different ethnicities that then get drawn into the study. We ask 'why do Koreans have a higher degree of stomach cancer?' and then compare across ethnicities. Our baseline is genetic differences." He argues that researchers must be attentive to the "economics of sample size and uniformity of data to give more cohesion," and in that sense they should already "know" or make which unknowns to be assembled. The official ethnic framing of racial medicine leads to research strategies that reproduce established ethnic hierarchies through data accumulation.

There are deep-seated beliefs in "relatively pure" genetic pools in "isolated" populations, wherein minority nationalities become potential stores of genomic resources for embattled Han bodies. There may well be political paranoia about what genetic benefits China's hardy, isolated minority nationalities harbor within them, to be mobilized to benefit a genetically deficient Han majority. Thus, the Tibetan DNA study mentioned above reveals how racial biomedicine is becoming a way of governing the near future, that is, within the realm of calculating cross-ethnic genetic benefits and weaknesses, set off against the backdrop of not only a complicated ethnic politics in China, but also the assumption of rapid environmental change at the planetary scale, with its concomitant effects on patterns of human living.

Adaptation to different ecological niches has caused "isolated" minority groups to develop adaptive genes in mountainous places while Han Chinese in other kinds of environments did not have to "hold on" to certain genes. Within this discourse of ethnic variability in acclimatization, comparative minority difference becomes a "recovery" of diluted genetic potentials. In addition, cli-

FIG. 9.5 BGI's Chinese name: "Greater China Genomics." Courtesy of Rena Lam.

mate change, which is also understood by researchers as "a near future," if not already being here, is inducing scenarios of understanding not only genetic relation but also genetic futurity, in the language of a physiological-genetic adaptation to globalized uncertainty.

Whereas, internationally, BGI promotes the sequencing of all life forms as a universal knowledge that it makes available to the world, the China-centered approach places value on genetic differences *between* human populations, especially those gathered by state nationalities' policy into the Chinese nationality (*zhonghua minzu*). Above, I discussed two BGI projects to capitalize on the value of genetic variation between ethnic groups: one to establish historical precedence of the Han, and the second to discover beneficial mutations that are unevenly distributed in order to develop new therapies for groups lacking the genetically beneficial traits. Genetic databases that seek to even out the uneven distribution of genetic adaptation between minority and majority groups express a biomedical topology of power. Here you have the emerging space spelled out in the Chinese name of BGI Genomics: "Greater China Genomics" (*Hua Da Jiyin*; see figure 9.5).

Shifting finally to a broader overview, my analysis has demonstrated contrastive Asian trends in the uses of ethnicity in biomedical sciences. BGI's approach in racial medicine may be called an "arboreal" or vertical modality, in contrast to Biopolis's rather more "rhizomatic" or lateral one.[32] Where insular Singapore's Asia is cosmopolitan and hemispheric in focus, continental China's scope is insular, in that it focuses on China's territory and the official diversity of its people as its site of investment. Han ethnicity is substantiated through projects like Yanhuang, which seems straightforward and expected. The bigger goal to build a national ethnic DNA cell bank appears to link human genetic diversity as moments in an evolutionary snapshot of iconic "immortalized cells." The genetic mapping of charismatic ethnic minorities and iconic "Chinese" species seems to reflect a way of scientifically establishing the broader genomic-national project, to think about the nation in terms of evolutionary divergences and continuities that bind the national unit *not* to a demographic latitudinal distribution across international space (as in Singapore) but, rather, to a kind of political longitudinal integration with other "Chinese" life forms within national space.

While BGI is an autonomous research institute, there are possible accommodations with officials, a shared sense of patriotism, and a sense that, outside the state system, BGI scientists can make better preparations for the nation's health uncertainties than the state sector itself. A BGI manager emailed me this from Shenzhen: "By 2020, we believe, [BGI] will be a critical player in life sciences development and pharmaceutical discovery." He projected the rise of a "distinctive model" of pragmatic collaboration among government research labs, top university researchers, and private firms that is expected to have more potential in pharmacogenomics than the 'Western' model where competing actors often work at cross purposes."[33] As a China-based and internationally oriented research institute, BGI contributes to the promises and uncertainties of genomic medicine. In Yang's words, it will "shake up" the world of cosmopolitan science.

A DNA BRIDGE AND AN OCTOPUS'S GARDEN

> The planet is moving through the void without any
> master. There it is, the unbearable lightness of being.
> —MILAN KUNDERA, *Milan Kundera:*
> *The Art of the Novel*

The DNA Bridge

Every day, thousands of people from across Asia visit world-class casinos in Macao and Singapore, gambling away millions of dollars on cards, dice, and games of chance. Not far away, in the laboratories of Singapore's Biopolis and China's BGI Genomics in Shenzhen, Asian scientists using the latest biomedical technologies are hard at work on a different set of odds—the game of life itself. These are the new Asian oddsmakers who have the power to remake the fortunes of billions of Asians.

In Singapore, the parallel games of changing fate are graphically configured in urban planning. On the waterfront, the stunning Marina Bay Sands Casino, composed of three terraced towers, is topped by a "skypark" that cradles an "infinity pool" (see figure E.1).[1] To some ethnic Chinese, the striking skyline looks like ancestors' tablets arrayed at an altar. The casino is connected by the Helix Bridge to the main island, where the Biopolis complex is situated. Designed by an Australian-Singapore consortium, the bridge is an architectural marvel that combines biotechnology and a yin-and-yang philosophy. According to the Singapore Tourist Board, the double-helical

FIG. E.1 The Helix Bridge connects a casino to the City of Life. Courtesy of the Singapore Tourism Board.

structure symbolizes values such as "life, continuity, renewal, everlasting abundance, and growth." The official reading is that the Helix Bridge brings "wealth, happiness, and prosperity,"[2] a favorite Chinese trio, to the new casino.

My view is that the DNA bridge and the ancestral-tablet casino, seemingly arranged in homage to kinsmen and staged and offset by one another, are each an image of descent and possible fortune on this threshold and microcosm of Asia. The overall architectural imagery is a projection, intended or not by their designers, of two versions of relatedness. As I have argued above, biotechnologies in Singapore have made explicit some of the underlying assumptions about kinship in Asia, even as modern science opens Singaporeans to new possibilities. More practically, the promised prosperity of the casino has been realized. Each day, millions of tax dollars generated by the casino are funneled into public coffers that, among other things, help to keep the Biopolis complex afloat. The new skyline is a hybrid symbol of taking and making bets: on money, on genes, and on an uncertain future.

Cosmopolitan science is in perpetual motion, in a state of vibration between fluidity and stabilization, or between forces of deterritorialization and reterritorialization.[3] Biomedical knowledge becomes territorialized in Asia through the application of codes that impose order on a profusion of differences, thereby stabilizing information in a particular situation. The main chapters have explored practices at Biopolis, where variations (in DNA, diseases, viruses, zones) are folded into one another to produce databases that, by stretching ethnicity, become applicable to majority populations in Asia. The last chapter shifted to BGI in China, where a national form of coding ethnic variations applies only to populations within the country. Whereas Biopolis deploys biomedical ethnic categories in order to make them generalizable across borders and thereby brands "Asian DNA," BGI identifies ethnic biomedical categories that are particular to China, intended for solving national problems of health and biosecurity.

In Singapore, life is made fungible by capitalizing on the analysis of ethnic differentiation—in sampling, databasing, clinical testing, and developing diagnostics—that seems necessary to the customization of "personalized medicine" for peoples in Asia. Although self-declared a commercial success, this frontier of biomedical science continues to confront different kinds of uncertainty—the continuation of state largess, a critical mass of experts, pharmaceutical investments, and scientific competitiveness—while building itself up as a biomedical center for confronting biothreats in Asia.

Singapore, which is already a major business platform, has built an infrastructure of international regulations in research and the life sciences. Biopolis is the center of a network that forms a zone of regulatory regimes within which cosmopolitan science can conduct experiments and gain access to resources and data on Asian human and nonhuman life forms. The state-driven endeavor is not limited to generating wealth per se, but a means of piecing together and leveraging various institutions to secure the island-state in a near future where disease looms large. Therefore, as a global biomedical platform for life-science research and marketization, Singapore is both a DNA bridge and an Asian satellite in the orbit of Western cosmopolitan science. It is a regulatory life-science haven in Asia eagerly sought by academic institutions and big pharma, outside of the unregulated research landscape in China.

What has been the impact of biomedical administration, jurisdiction, and subjectivation in shaping affective engagement in the Singapore public? Aside from the flurry of comments surrounding the formation of the Bioethics Advisory Committee, there have been no significant public comments other than concern that medical tourism may cut into local medical care. Only a playwright, Ong Keng Sen (not related to the author), has tried in Singapore to pose ethical questions about the biomedical initiative. On a website for his play "Facing Goya," Ong explains that "the opera asks what are the ethical limits of science, medicine and biotechnology? Can creativity be cloned—should it—and is there a price to pay as nature is gradually straitjacketed by scientific advancements?"[4] As an artist in the humanistic vein, the playwright, by criticizing the threat that modern science poses to creativity as a radically subjective agency, does not surprise here, but his point about being straitjacketed is a challenge to the Singaporean presumption that scientists are morally authorized to make decisions for the common good until proven wrong, thus obviating the need for an informed public discussion.

Sophisticated science demands an engaged public equal to its ambition to change the forms and norms of life as we know it. Drama is a good place to begin, but playwrights should be joined by social scientists and philosophers to start a public conversation about the wide-ranging implications of experiments with life in their midst. What are the benefits and uncertainties of sequencing DNA and perturbing cells as lifesaving therapies in Asia? How can the public be party to discussions on ethical limits to such experiments? Like any best-conceived biopolitical plans—to improve health and wealth—the innovative ethnogenomic experiments in Singapore can go awry. Bioinformation and cell lines are scientific enactments and productions of life, and thus far from being life actually lived, if such a distinction still retains the shimmering clarity of certainty.

This book has been about how contemporary, experimental life sciences have generated new practices of modeling, and thus emergent fields of knowledge of life. We are entering a period when the algorithmic approach is assumed to exceed the human grasp of information, and computerized intelligence complements the biological sciences, thus posing a threat to older knowledges of vitalism. Among the questions raised in Asia will be the impact of biomedical science on rich healing traditions: Ayurvedic, Chinese, Tibetan, Malay, Thai, Vietnamese, among others. The ineffable power of traditional Chinese medicine (TCM) continues to exert its presence throughout the region and beyond. Anthropologists have identified multiple "traditional" medical practices—herbal medicine, food as therapy, acupuncture, regulation of energy flows to achieve balance and harmony, and a focus on everyday practices of self-diagnosis and self-regulation—as compelling features capable of generating health and particular forms of life.[5] In Singapore and beyond, TCM continues to flourish through healers, herbalists, acupuncturists, masseurs, and tai chi and qi gong masters, whose ways of knowing and practicing the human body remain an art—at once humanely impersonal and yet intensely personal.

By contrast, biomedicine depends on scientific experiments and finding a finer fit between molecular processes and novel drugs, bypassing "healing narratives" in favor of personalized therapies focused not on the prevention of disease but its intervention.[6] At the same time, there are eerie convergences of TCM and biomedicine. The body as a self-healing system through the unblocking of energy flows in acupuncture is echoed in molecular techniques to unblock checkpoints in immunotherapy. These remain highly distinct systems of health making, cultural versus technological, and are alternative approaches that exert different subjectivizing effects on the patient. For decades now, peoples in Asia have availed themselves of both "traditional" healing practices and modern high-tech medicine, but health politics have great implications beyond the individual.

Notwithstanding pride in ancient healing wisdom, there are high national and geopolitical stakes in deploying the latest bioscience and medical technologies for the new mappings of things human in Asia. The respective enigmatic machines devised in Singapore and in Shenzhen and Hong Kong, in their particular ways, reanimate other kinds of cultural interests not only of the modeling of data but also of the circulation of concepts, desires, and projections beyond the region.

The Octopus's Garden

Besides a superficial similarity in the use of the ethnic heuristic, BGI, China's foremost bioscience institute, is a rather different entrepreneurial science model than Biopolis. As an infrastructural powerhouse, it has the capacity to make different rules in research and to play a role in shaping cosmopolitan science in the present-future.

Whereas Biopolis is a state initiative, BGI Genomics is a private, "citizen-managed" (min-yong) company operated by independent, China-born entrepreneurial scientists. But because it is impossible to be a big company in China without some form of state involvement, some officials are probably major investors in BGI. It claims to have the highest concentration of PhDs (two thousand) and other highly trained research staff, including two thousand biostatisticians alone, of any science facility in the world. The center's formidable sequencing capacities have become a springboard for new drug discoveries and advanced genetic research. As a global company, BGI has the capacity to transform public health policy worldwide because it enables mass affordability for sequencing an individual's DNA to personalize his or her medical treatment. From its China base, BGI has actively sought partnerships with universities and hospitals in Europe and the United States, and with the Gates Foundation. It has sought to become a new model for the Chinese company. The combination of its vast DNA sequencing and data-mining powers, the potential information of over a billion Chinese people, and a relatively loose regulatory environment in China have spurred its fast ascent in the world of genomic science that rivals centers in the West.

By comparison, at Biopolis, biomedical discourses invoke "Asia" as a multiple and mutating territory of biomedical governance. Singapore's strategy is to be global, by deploying Anglicized ethnic terms as simplified but stable biomedical categories that may be stretched across borders and applicable to "Asian" peoples at large. By contrast, at BGI, "greater China" is the projection of the company's global biomedical ambitions. If considered together (though they are institutionally distinct entities) both hubs, variously wielding the demographic, biomedical, and biotech weightiness of "Chineseness," suggest the formation of emerging cosmopolitan science at once specific to Asia and applicable through the world. But whereas Singapore is a science milieu in Asia regulated by rules and regulations and commensurable with those in Euro-America, BGI operates in a poorly regulated science environment in China and thus is better able to use "Asia" (China's scale, population heft, and inexpensive high-skilled labor) to leapfrog ahead as an innovative global science enterprise.

When it comes to biomedical ventures coming out of China, there is great anxiety about the modeling science power, motives, and curation—in the sense of caring, organizing, and taking charge—of life forms represented by BGI Genomics. East Asian scientists have been stereotyped not as hypothesis-driven or visionary and innovative in their research, but as having an industrial mentality of seeking optimal outputs via labor-intensive mass production akin to the manufacturing of consumer products. There is foreign concern as well that Asian researchers may be disinterested in the kinds of virtue that have long defined the scientific vocation in the West, when many are caught up with questions of how biomedical science can benefit their nations as well as peoples in the developing world. Thus BGI, for its Chineseness, is a more elusive research entity than Biopolis, and it is viewed with ambivalence and even fear by observers in the West.

As an octopus-like research institute with tentacles in different businesses and institutions at home and overseas, BGI is perhaps more comparable to the J. Craig Venter Institute, which integrates different organizations (including Synthetic Genomics) related to genomic research, energy, and synthetic biology. It would be a mistake, however, to confuse the Venter-esque and BGI models of scientific entrepreneurialism. BGI is a global company with a massive infrastructure and an assemblage of scientific brainpower that costs much less than in Singapore or the United States. The BGI monopoly on sequencing power has stirred trepidation in the science world, about whether the absolute scale of the BGI production of data, the relatively inexpensive science workers, and the potential database containing details about 1.3 billion people in China would give it a competitive advantage in biosciences. BGI is indeed a new model whereby the unregulated science milieu in China allows for the speed and scale at which the science behemoth is the first global Chinese science company, with the potential to shape the future of the life sciences.

In an interview addressing Westerners' worry about the nature of the institute, Wang Jian responded that making money was the way to fund basic research in order to "do good." He compared BGI's strategy to that of "a wandering migrant worker—looking for opportunity and occasionally irritating the authorities."[7] As a global science institute, BGI is realigning different research, health, and business institutions in the world through its command of sequencing infrastructure. Whereas the Venter Institute reflects the philosophy of its flamboyant founder as a "go-it-alone" science entrepreneur, BGI's model is to launch a new People's Republic of China (PRC) brand of entrepreneurial science operating simultaneously on global and domestic fronts.

Indeed, BGI officials regularly held up Venter and his use of synthetic biology to create novel living forms as models *not* to emulate. Here they articulate an alternate notion of virtue in science, that is, not to be driven by scientific innovation for its own sake or propelled by the yen to make controversial discoveries (although it would be nice to achieve global recognition for their work). Dr. Chen of BGI suggested that, as socialist pragmatists, he and his colleagues are preoccupied by the kinds of research that can lead to affordable biomedical products that can eventually be made available to the masses. BGI can bring the bold scale of its operations to bear on Chinese national problems of long-term food and health security. The implication as well is that BGI can do what the Chinese state cannot do on its own, which is to apply genomic science as a preparation technology to meet rising challenges of health, food, and even energy needs. BGI's *minzu*-differentiated data are concerned with populations boxed within China's official ethnic classification system, in contrast to Biopolis's use of widely circulating ethnic designations inherited from colonial rule that now function as a mobile heuristic that allows data to move and flow. Therefore, despite the suave cosmopolitanism of its founders, BGI racial analytics recycle *minzu* categories, giving new meaning to politico-aesthetic differences.

One obsession of postmarket-reform China has been the human quality (*suzhi*) of its vast population, especially when differentiated by class in relation to global capitalism.[8] The moral devaluation of physical labor is paralleled by a valorization of *wenhua* (culture) now inseparable from the acquisition of human capital. What is the role of the life sciences to securing this human capital from the genetic and demographic scale of the Chinese dragon? Leading Western-educated scientists have come to embrace the American notion of the "technological fix" for China's myriad problems that all boil down to the human biological of the present-future. Here is a BGI response in 2012 to my query about the importance of genomic science to China:

> Genomic and inheritable diseases are very severe in China, which can be shown in the following data. In China, one patient dies from leukemia in every 12 minutes and one woman dies from cervical cancer in every 6 minutes. And patients with Down syndrome have an average occurrence of newborn patient [*sic*] in every 20 minutes in China. At present, there are more than 600,000 patients with Down syndrome, with an incidence rate from 1/800 to 1/600. There is also a high incidence rate of thalassemia in the Guangdong Province, Guangxi Province, and Hainan Province. About 3000 babies with thalassemia are born every year. Therefore, accurate, safe and non-invasive genetic testing for genomic and inheritable diseases is very crucial for the public. For example, non-invasive Prenatal Genetic Testing of Fetal Chromosomal Aneuploidy only requires 5ml maternal peripheral blood and the intrauterine infection of the fetus, and miscarriage can be avoided. Genetic testing on inheritable diseases can greatly relieve the burdens of families and society.[9]

But there is the chance that this highly unregulated science milieu of well-intended biomedical interventions may spin inexorably into unnerving experiments. As Susan Greenhalgh has argued, in China's modern quest to escape backwardness, the single-minded PRC pursuit of science to solve problems of overpopulation has inflicted suffering on millions of Chinese through its family planning program, one that promotes a curation of humanity that is troubling.[10] It is important to note that BGI is a nonstate, global company, and it should not be confused with the Chinese government. But not surprisingly, capitalizing on Chinese demographics, BGI also has its own projects for "improving" and curating human life.

BGI's entry into the field of cognitive genetics is marked by a focus on finding the genetics of brain disorders and intelligence (IQ measures), and acting on such data to benefit parents. BGI is a leading member of an international autism consortium that includes the Duke University School of Medicine. BGI's sequencing prowess in whole genome sequencing and analysis has identified de novo or rare inherited mutations that give rise to autism in Autism Spectrum Disorder groups. Commercially, BGI has produced a noninvasive test (the above prenatal genetics test of fetal chromosomes) called NIFTY that scans the pregnant woman's blood for signs of autism in the child she is carrying. Worldwide, NIFTY is one of the safest, most reliable, and affordable tests for autism in unborn children.

A related and more controversial BGI project is the engineering of "genius babies" through zygote selection before implantation in the mother. In 2012, BGI scanned the DNA of two thousand brilliant individuals chosen according to the problematic IQ measures, with the goal of finding how genetic variations affect intelligence. Half of the sample are adult Westerners such as professors, and half are child prodigies from China's best high schools. The findings allow BGI researchers to screen weeks-old human embryos for so-called IQ genes. The technology is a more intricate technique than the widespread use of ultrasound to screen for the sex of a baby. Testing and ranking zygotes for intelligence genes in a Petri dish allow parents to select an embryo in a sibling cluster that promises to have the highest IQ points for implantation and birth. Ethical issues about culling embryos for the purpose of having the smartest babies do not exist in the PRC. Indeed, IQ tests for zygotes may be the very thing that many people want for family planning.

Here we see an example of how the latest genetic technology allows a recurrence of cultural objects that come to take on new meanings. The chauvinist claim of a "five-thousand-year-old civilization"[11] includes beliefs in the historical inheritance of superior intelligence in "Chinese people." The demographic and market strategies of an ascendant PRC further entrench the worship of brainpower. Decades of draconian family planning that limits majority Han Chinese to "just one child"[12] have raised the stakes of having a smart kid who can struggle in a densely populated and ferociously competitive country. An American psychologist who contributed his DNA to the genius project observed, partly in self-defense: "In China, 95 percent of an audience would say, 'Obviously, you should make babies genetically healthier, happier, and brighter!' There's a big cultural difference."[13] The technology, which is a kind of familial eugenics—private, assisted selection of embryos also happens in advanced liberal countries—can open the way to genetic engineering of humans in the future. Because studies linking genes to intelligence are largely taboo in Europe and the United States, BGI can gain a head-start.

Nevertheless, in other ways, BGI as a research milieu is definitely more regulated and international than other biomedical research sites in China. In 2015, scientists at Sun Yat-sen University in Guangzhou used the Crispr method (invented by scientists at MIT and Berkeley) to edit genes within defective human embryos. The editing of genes to cut out disease-causing genes in a fertilized egg can inadvertently induce other mutations, therefore making this an ethically charged procedure. The Guangzhou experiment stirred an uproar across the Pacific, with warnings from one of the inventors, Dr. Jennifer Doudna of Berkeley, about using Crispr therapeutics for genetic manipulation.[14] Because they used defective embryos, scientists at Sun Yat-sen University were trying out a step and not trying to mutate normal embryos. But clearly this was a first step in an experimental quest to eventually edit genes that cause beta thalassemia, a serious blood disorder (mentioned in the BGI statement above) that is prevalent in South China.

The Unknowns

Dr. Ma, the stem cell scientist at Biopolis, observed, "We are in a period of fast science. The only way to know where you are going is to look back." But while capturing particular sets of presumed "known unknowns," scientific practices are proliferating a swarm of "unknown knowns." Slavoj Žižek warns about the "silent presuppositions we are not aware of, [that] determine our acts." He continues by asking whether contemporary trends in digitalization and biogenetic manipulation "open themselves up to a multitude of options?"[15] We are reminded of the "butterfly effect," or the initial tiny wing flutter someplace that sets off a typhoon from afar.

This book has illuminated how biomedical technologies are variously deployed to manage health problems in the Asian present-future. The first strategy is to amass risk calculations on health susceptibilities by ethnicity and disease in order to customize medicine for peoples in the region. A second set of challenges identifies not-so-easily-managed risks, in that uncertainties surround the "success" of biomedical endeavors. Despite the carefully assembled Biopolis project, the question of long-term sustainability haunts their particular approach to biomedical entrepreneurialism, and the costs thereby entailed, the capacity to accumulate a critical mass of science experts, and the promise of certain stem cell experiments. Then there are those uncertainties I call the known unknowns, referring to the combination of anticipated catastrophic events and the uncertain capacities to prepare for managing them. Here science confronts possibilities shaped by culture, politics, and nature. Political and cultural tensions between Asian nations threaten to undermine embryonic efforts to build a sense of scientific community in the region. Despite Singapore's preparations to be the biosentinel hub for a region beset by newly emerging infectious diseases, cross-border health collaborations are still weak and perhaps unreliable. The rise of Biopolis and BGI Genomics also stirs known unknowns surrounding cosmopolitan science itself.

As an extension of Euro-American sciences, Biopolis strategically deploys U.S.-generated medical concepts and therapeutics to reposition "Asian" racial categories as abstract, flexible, and mobile categories for sampling, diagnosis, and customized medicine. Once formed, the expansive lab, the genomic atlas, and the ethnic immutable mobile mobilizing "Asian" biological vulnerabilities all configure a kind of geo-biosocial ethics that makes moral demands on scientists to anticipate any external threats—rising rates of aging diseases and newly emerging infectious diseases, pressures to feed huge populations—in the Southeast Asian ecosystem.

Scientific entrepreneurialism is about grappling directly with uncertainty. As a city-state balanced on the edge of precarity, Singapore generates an ethos of *kiasu*, an orienting fear of "losing out" that also demands risk-taking, if only to avoid catastrophes. The turn to the life sciences forges an interesting convergence of the Singaporean fear of losing out and the U.S. fear of failing in the Asia arena. Biopolis as a biomedical platform knits together American expertise and Asian materials, thus becoming a frontier for expanding U.S. scientific entrepreneurialism into the region.

By comparison, BGI's mode of scientific entrepreneurialism is exercised mainly on its own terms. The *minzu* nationality categories used in BGI are biological manifestations of official ethnicities, known in advance only to be "discovered" again. Post-genomic technologies corralling genetic biodiversity in humans, animals, plants, fungi, and microbes are a vital part of Chinese order. They are critical new tools for a giant nation newly able to deal with the modern issue of biosecurity. The life sciences in the PRC pose the question of whether biomedicine can taxonomize societies and the world in new ways. The uses of biomedical technologies that specify persistent health threats and benefits in nationally framed ethnic terms configure a space of intervention that may give rise to "the birth of immunopolitics."[16] At what point will a strategy of immunizing ethnicized bodies from ethnic-specified diseases lead to the immunization of the body politic from external threats to China's strength? Does the Euro-American approach to post-genomics provide a real alternative capable of actualizing the promise of a life (borderless, near-future) that would be worth living? Such are the questions about the not-yet-knowable.

The future leadership and direction of cosmopolitan science, necessarily involving multiple institutions and actors, are also in question. Scientists and other observers in Europe and America are already noting the sheer scale and unregulated acceleration of research in China. One BGI official confided that, in the past, Chinese scientists were centuries or at least decades behind modern sciences originating in the West. Scientists from Asia suffered from the lag time as they sought to catch up with the latest knowledge, discoveries, and innovations in the physical and life sciences, especially in chemistry, physics, astronomy, and biology. But today, with tens of thousands of Chinese scientists enrolled in research institutes in Europe and the United States, and the rapid modernization of universities in China, Dr. Chen of BGI said, "We are starting off from the same baseline, especially in the field of postgenomics." As I have indicated in this book, besides China, elsewhere in the Asia-Pacific—Japan, South Korea, Taiwan, Singapore, India, as well as Australia and New Zealand—researchers are not only catching up but also competing on newly equal terms with individuals and institutions in Europe and the United States, and in some ways, setting new terms for regional and intraregional practices of cosmopolitan science. And because many scientists in China operate in milieus lacking formal ethical constraints, they may pull ahead in certain highly controversial areas. The fearful thing about China's new class of experts is that they may be not sufficiently afraid of failing.

For all these reasons, BGI has a very important role in mediating between cosmopolitan (regulated) science and unregulated science in China. It is important to note, however, that failures in ethical considerations also happen in the heartland for cosmopolitan science. For instance, in 2012, Tufts University launched an investigation into ethical violations in an experiment that fed genetically modified ("golden") rice to Chinese children. There was a lack of coordination between review boards at Tufts and those in Chinese institutions where experimental subjects were recruited.[17] In other words, efforts to sustain best practices in cosmopolitan science require constant vigilance in universities and research institutions in North Atlantic countries, the originary home of international science, as well as in emerging global centers.

Perhaps just as Biopolis spreads reputable science practices throughout Asia, BGI can be a disseminator of ethical "best practices" and regulatory regimes in research institutions in the PRC. Because BGI has octopus-like connections to Euro-American science, and a sincere desire to do good for the world, it is a major player with the weight of the world's most populous nation behind it. As a China-based, infrastructural powerhouse, BGI may be the strategic site to grow and disseminate ethical guidelines for China's numerous scientific organizations. There are many unknown unknowns, but it is hard to see the elephantine, corruption-ridden, and chaotic Chinese state in the driver's seat when it comes to the life sciences. Innovative Chinese scientists at home and in the diaspora, as are participants in cosmopolitan science, seem poised to play a role in the science of the future. What role, what modality of science, and what futures however, remain, to be seen.

Fortune, Fungibility, Fear, and Finitude

My admittedly partial accounts of Biopolis, and of BGI Genomics, in different ways pose the question of the future of cosmopolitan science. As biotechnologies enhance our capacity to make life fungible, and Asian research frontiers discover, endow, and deploy novel epistemic objects with complex meanings and affects, what kinds of fortunes are being pursued and, perforce, found?

A major theme has been the capacity of researchers at Biopolis to make more of life, in part by folding data points linking DNA variants, ethnicities, diseases, and even geographies into flexible and game-changing databases. At the same time, BGI's bioinformatics infrastructure operates at a scale so bold that it promises to draw major human, animal, and plant species, including the octopus and the bamboo, into its folds. These enigmatic machines, one perturbed by the affects of anxiety and optimism, the other by the affective effects of PRC science order, may well have—in their respective ways, very differently from J. Craig Venter's ventures—shifted the very notion of "progress" in science.

In Biopolis and BGI Genomics, it is no longer about Asia being left behind in the march of scientific progress or being drowned in a hostile regional neighborhood. Rather, both cases demonstrate that the biomedical sciences indeed create their own ways of envisioning the scales at which progress can be defined and strived for. Evolving biomedical science in Asia, as analyzed above, seems to be in a different orientation than the Faustian bargain of Venter-esque science (a high-philosophical, risk-taking, big-stakes sacrifice). Rather, researchers in Asia are less interested in big claims and world-changing breakthroughs than in the realization of small fixes that nonetheless reorganize the Asian present and its concomitant projects and projections of Asian modernity. In this as well, the ethnic heuristic has been pluripotent in generating affective validation in successes gained so far.

Thus, an overarching claim of this book is that what is at stake is not high science that charts the infinite terrain of progress. Rather, situated science projections and enactments of an anticipatory future depend on a re-assemblage of disparate components and moments through which the present and the future's relationship is reorganized. This study has found resonances between the operations and conceptual infrastructures of cells, biomedical research, and biological processes on the one hand, and the structuring of research institutions and the ethnographer's theoretical and conceptual work on the other. The mechanistic philosophy that sees the body as a well-calibrated machine is distinct from the inner workings of a cell, which stand as a reminder that processes of all kinds are often not teleologically structured toward any guarantees. Let us pause a moment and consider a lesson from cellular biology.

One reason that biologists investigate the microscopic intricacy of life is that the cell is more than a tiny factory or waste disposal entity. The cell has a genome, pathways, and proteins. But the cell must have sufficient capacity to fold proteins into their appropriate three-dimensional shapes. The failure to fold protein, a microscopic origami within the cell, has been linked to human diseases as diverse as neuro-degeneration and cancer. A cell can synthesize, in the origami of survival and synthesis, proteins that not only fail but also irrevocably alter and eventually destroy the (organic) systems that create and sustain them.

The cell powerfully resonates with the ways in which the microscopic and the regional interact in biomedical milieus and in this anthropological study. For instance, is Biopolis the cell or is it the protein, as it itself folds disciplines, samples, and information into each other? At the same time, is the Biopolis complex/ecosystem also subject to foldings of different kinds (pharmaceutical markets, science rivalry, geopolitical stakes, affective [dis]trust, etc.)? As a conceptual metaphor, the protein origami also suggests the vertiginous openings and collapsings of scale that this ethnography has hoped to capture.

Another cellular lesson would be that the biomedical assemblage, by means of the momentum accrued by the force of its interlocking relations, may stall or spin out into forms that are virulent or toxic. Can scientists who operate under specific regimes of virtue remain, nonetheless, somehow blind to the possibility that the enterprise itself might create malignancies, or failed and yet still consequential protein foldings as it were? For instance, researchers may fold their many data points, linked across too many dimensions, incorrectly and to great consequence. There is, depending on what register is involved, a strong probability or likely a kind of spurious guarantee of fungibility and flexibility triggering an overvaluation of aggregated datasets. With the digital machinery of life, a synthesis of multiplier effects, and data flows having different points of crystallization, will the cascade of minutely aligned information collapse and become meaningless?

In the anticipatory, calculable order of Asian science, one should not disregard the haunting affects of fear and hope, doubt and precariousness. The refolding of present and past (biomedical and "uncivilized") bodies becomes part of an extended human toolkit of surrogates. In both BGI and Biopolis, there is an accelerated capacity to engineer surrogate future-bodies that fundamentally change the meaning of human corporeal boundaries. The implications of surrogate alter-selves are different than robots replacing human workers. The future of biomedicine, we are told, rests on immunology, and the valorization of the actual (privileged, dominant, wealthy) patient means scientists are tempted to find living substitutes. For instance, the BGI discovery of the "athlete gene" depended on minority Tibetans becoming biomedical surrogates for the majority Han Chinese. As scientists focus on designing the immune system as a therapeutic platform, our extracted organs come to act as lab surrogates for our humanized bodies. New modes of biomedical inequalities may be produced among different kinds of living forms. What do trends of using human and organ surrogates portend for novel efforts to secure the near future of Asian bodies and life?

There is the danger as well that novel assemblages of biomedical innovations may be destabilized by affective agitations, or they may inadvertently trigger a series of interconnected risks that spill across populations, markets, ecologies, and regions. The "order of things," Foucault reminds us, bumps up against the finitudes of life.[18] For scientists in Asia, questions of intervention, improvement, and security, rather than any high-minded scientific virtue, are at stake. Their experiments are about making ways of living and making more of life, creating the techniques, infrastructures, and therapies to sustain an Asia through coming contemporary challenges. But even in this multitudinous tinkering, human mastery of nature remains in doubt, and scientists realize the enigmas that haunt their pursuits, as well as the uncertain nature of self and experiment that Milan Kundera calls the "unbearable lightness of being." The life sciences, and science in general, would do well to engage their existentialist challenges, if not finitudes.

Prologue

1 23andMe solicitation letter signed by A. Wojcicki, September 30, 2013. Copy on file with author.

2 When the semester began, some professors objected to the experiment on grounds of its violation of student privacy and for lacking an educational briefing on personalized medicine. The data thus gathered from incoming freshmen was subsequently anonymized (for research?) and posted on a website forum. One finding (not new) is that many students (e.g., Asian Americans) have a genetic tendency to be lactose intolerant.

3 The Genetic Information Nondiscrimination Act was passed in 2008, but personal DNA information is commercially available.

4 By 2015, 23andMe decided to strengthen its health component by teaming up with drug company Genentech to use the database for finding therapeutic targets for Parkinson's disease.

5 In actuality, the National Institute of Health (NIH) has created a new American model of socially robust medicine that links academic research in drug discovery and state investments in public infrastructure and commercial companies. A new 2015 "precision medicine initiative" will provide funds for collecting the genetic data from one million American patients.

6 See Guyer, "Prophesy and the Near Future," and Fischer, *Anthropological Futures*.

7 "Enigma" was the name of the coding machine used by Germans during the Second World War for enciphering and deciphering secret messages.

8 Traweek, *Beamtimes and Lifetimes*.

9 Lock, *Encounters with Aging*, and *Twice Dead*.

10 Kleinman, *Patients and Healers in the Context of Culture*.

11 Cohen, *No Aging in India*.

12 Sunder Rajan, *Biocapital*.

13 Cohen, "Operability, Bioavailability, and Exception"; Scheper-Hughes, "The Last Commodity"; Wilson, "Medical Tourism in Thailand"; and Vora, *Life Support*.

14 Ong and Chen, *Asian Biotech.*

15 Collier and Ong, "Global Assemblages, Anthropological Problems."

16 Shapin, *The Scientific Life*, 5.

17 "Irrational exuberance" was coined by Alan Greenspan, the former U.S. Federal Reserve chairman, to describe market speculations that led to crises.

18 See Foucault, *Knowledge/Power*, 204.

19 See, e.g., Washington, *Medical Apartheid.*

20 Besides Washington, *Medical Apartheid*, see, e.g., Montoya, *Making the Mexican Diabetic*; Wailoo, *How Cancer Crosses the Color Line*; and Pollock, *Medicating Race.*

21 Duster, *Backdoor to Eugenics.*

22 Lock and Nguyen, *The Anthropology of Biomedicine*, 353.

23 Duana Fulwilley, "The Molecularization of Race: Institutionalizing Human Differences in Pharmacogenetics Practice," *Science as Culture* 16, no. 1 (2007): 1–30.

24 Lock and Nguyen, *The Anthropology of Biomedicine*, 353–358.

25 Hacking, "Making Up People."

26 Nelson, "DNA Ethnicity as Black Social Action?"

27 Bateson, *Steps to an Ecology of Mind*, 318.

28 Luhmann, *Observations on Modernity*, 16, 48.

29 Charis Thompson, "Race Science," *Theory, Culture & Society* (special issue on problematizing global knowledge) 23 nos. 2–3 (2006): 547–549.

30 Hartigan, "Mexican Genomics and the Roots of Racial Thinking."

31 Wade, Beltran, Restrepo, and Santos, *Mestizo Genomics.*

32 Deister, "Laboratory Life of the Mexican Mestizo." For the notion of genes as national cultural patrimony, see Rabinow, *French DNA.*

33 Chua, "The Cost of Membership in Ascribed Community."

34 Yang, "The Stats behind the Medical Science," 22.

35 Collier and Ong, *Global Assemblages.*

36 See Fischer, *Anthropological Futures*; and Marcus, "The End(s) of Ethnography," 1, 12.

37 Rose, *The Politics of Life Itself*, 183, 86.

38 Rose, *The Politics of Life Itself*, 29.

Introduction

1 George Yeo, "Singapore Must Be Like an Italian Renaissance City-State," 938 LIVE radio station report, August 13, 2007.

2 In the lead is Helios/Titan, the personification of the sun, who drives the chariot each day through Oceanus (the equatorial flow) and returns to the east at night. Chromos, a brief form of chromolithograph, identifies the techniques of vision and color itself, thus denoting the saturation of hues or "value" in art. Nanos refers to microscopic scales of space and time. Matrix in zoology refers to the generative part of the animal, thus referring to origins and to the milieu of productivity in a living system. Genome identifies that which produces, and in Greek it is also the stem for "race" or "offspring." And Proteus is a sea god who can change his form at will, symbolizing mutability in biology. These Greek terms have been liberally applied to science artifacts and techniques. Thanks to Gabriel Coren for rendering the symbolisms clearer to me.

3 Quoted in G. Traufetter, "Biotech in Singapore: A Treasure Island for Elite Researchers," *Der Spiegel*, March 26, 2005, http://www.spiegel.de/international /spiegel/0,1518,349122,00.html.

4 World Bank, "GDP per Capita by Country," http://data.worldbank.org/indicator /NY.GDP.PCAP.CD?order=wbapi_data_value_2013+wbapi_data_value+wbapi _data_value-last&sort=asc, accessed March 28, 2015.

5 Orville Schell, "Lee Kuan Yew, the Man Who Remade Asia," *Wall Street Journal*, March 27, 2015.

6 In Mandarin Chinese, *kiasu* is *pashu*.

7 Chua, "Singapore as Model."

8 Clancey, "Intelligent Island to Biopolis."

9 Sydney Brenner is a leading pioneer in genetics and molecular biology and a fellow of the Crick-Jacobs Center in the United States. Early in his career, he worked with Francis Crick at Cambridge. Brenner's work contributed to the analysis of gene sequencing, and he established the existence of messenger RNA (ribonucleic acid), which functions as a blueprint for the genetic code. In 1996, he founded the Molecular Sciences Institute in Berkeley, California. Besides receiving recognition from Singapore, Brenner has received awards from BGI Genomics in Shenzhen, China.

10 See Ong and Chen, *Asian Biotech*, 21.

11 Sunder Rajan, *Biocapital*.

12 Hayden, *When Nature Goes Public*.

13 See Ong, "Scales of Exception."

14 For originary conceptions, see Marx, *Capital*, and Benjamin, *Work of Art in the Age of Its Technological Reproducibility*. Classic anthropological works include Mauss, *The Gift*, and Malinowski, *Argonauts of the Western Pacific*. More recent explorations of the value-laden nature of economic activities include the symbolic meanings generated by the Atlantic sugar trade (Mintz, *Sweetness and Power*); the "marginal gains" of African money (Guyer, *Marginal Gains*); and the "everyday communism" that underpins traditional credit systems (Graeber, *Debt*).

15 Helmreich, *Alien Ocean*.

16 Kay, *Molecular Vision of Life*.

17 Jasanoff, *Designs on Nature*.

18 Rheinberger, *Epistemology of the Concrete*, 10.

19 See Rheinberger, *Epistemology of the Concrete*; Kay, *Molecular Vision of Life*; Jasanoff, *Designs on Nature*; Haraway, *Simians, Cyborgs, and Women*; Haraway, *Modest_Witness @Second_Millennium*; Rabinow, *French DNA*; Rose, *Politics of Life Itself*; and Clarke, Shim, Mamo, Fosket, and Fishman, "Biomedicalization."

20 Collier and Ong, "Global Assemblages, Anthropological Problems."

21 HUGO was inspired by Sydney Brenner. The international organization was founded in 1988, in Cold Spring Harbor, New York, and it has members from two dozen countries. HUGO focuses on the study of human genetic variation in relation to environment, disease, and treatment. See the Human Genome Organization website, "History," http://www.hugo-international.org/HUGO-History, accessed June 5, 2015.

22 For an account of scientific entrepreneurialism in American industry, see Shapin, *Scientific Life*, 209–267.

23 Ong and Chen, *Asian Biotech.*

24 Weber, "Science as a Vocation."

25 See Fortun, *Promising Genomics*, 11.

26 According to an A*STAR (2015) tenth-anniversary report, Biopolis has "put Singapore on the scientific world map for biomedical research." Although Biopolis began with the state as a venture capitalist, the trend has been to promote public-private collaborations. Its research strategy to develop ethnic-stratified medicine has attracted many investments in research initiatives and made Singapore a venue for clinical tests for foreign companies. A biomedical manufacturing sector comprising drug companies scattered across the island—Genentech, Merck, GlaxoSmithKline, Sandoz, and Novartis, to name a few—contributed close to $30 billion in revenue in 2012. Besides the value-added contribution of the drug and biotechnology industry, scientists in Singapore have advanced the understanding of cancer, eye disease, neuroscience, metabolic diseases, and infectious diseases. A detection kit for SARS and a vaccine for the H_1N_1 flu virus are among Biopolis's achievements. The official "success story" therefore depicts a more varied picture than narrow commercial goals from the kind of late capitalist, big pharma–funded contexts of innovative science in the United States.

27 Shapin, *Scientific Life.*

28 Ong and Chen, *Asian Biotech*, 38–39, 89–91.

29 Foucault, *Security, Territory, Population*, 11.

30 Guyer, "Prophesy and the Near Future."

31 Foucault, *Security, Territory, Population*, 70–71.

32 Rose, *Politics of Life Itself*, 17–18.

33 Buchanan, "Deleuze and Geophilosophy."

34 A*STAR Research, "Singapore's Biopolis: A Success Story."

35 Waldby, "Stem Cells, Tissue Cultures, and the Production of Biovalue," 310.

36 Waldby and Mitchell, *Tissue Economies.* See also Rose and Novas, "Biological Citizenship," 455.

37 See Moor and Lury, "Making and Measuring Value," 451–452.

38 Bateson, *Steps to an Ecology of Mind*, 318, 400. See also Luhmann, *Observations on Modernity*, who identifies the autopoetic system as operating within a larger ecological context.

39 Deleuze, *The Fold.*

40 Bowker, "Biodiversity Datadiversity," 643.

41 Callon, "What Does It Mean to Say That Economics Is Performative?"

42 See, e.g., Rose and Novas, "Biological Citizenship."

43 Law and Mol, "Situating Technoscience."

44 Luhmann, *Observations on Modernity*, 95.

45 Lakoff and Collier, *Biosecurity Interventions*; Lakoff, "Preparing for the Next Emergency," 247.

46 Beck, *Risk Society.*

47 In earlier times, the division between the "normal" and the "pathological" in the life sciences was fundamentally a moral exercise that established adaptation to a particular standard of social normativity. See Canguilheim, *Knowledge of Life*, 121–133. But

in contemporary biopolitics, the relation of the standard to the norm has shifted as political rationality and scientific rationality are increasingly entangled.

48 Foucault, *Security, Territory, Population.*

49 Desrosieres, *Politics of Large Numbers.*

50 Deleuze and Guattari, *Thousand Plateaus.*

51 Lakoff, "Diagnostic Liquidity," *Theory and Society* 34, no. 1 (2005): 67–68.

52 See Keating and Cambrosio, *Biomedical Platforms.* Besides infrastructural development, the shift from medicine to biomedicine involves redefinitions of the human body, diseases, and therapies. See also Clarke et al., "Biomedicalization."

53 Quoted in Errol Morris, "The Certainty of Donald Rumsfeld, Part III," *New York Times,* March 28, 2014, http://www.realcleardefense.com/2014/03/28/the_certainty_of_donald_rumsfeld_part_iii_260988.html, accessed December 12, 2014. Rumsfeld was responding to journalists who were seeking information about how the Pentagon established a link between Saddam Hussein's regime in Iraq and weapons of mass destruction. Journalists found that Rumsfeld's use of the term "known unknowns" for the imponderables that are part of military decision making was an evasive ploy to avoid telling the truth.

54 Morton, *Hyperobjects.* "Known unknown" risks are less improbable than the black swan events (e.g., the 9/11 terrorist attacks on the United States) identified by Nassim N. Taleb in *The Black Swan.*

55 See also Lakoff, "Preparing for the Next Emergency"; Samimian-Darash, "Governing Future Potential Biothreats."

Chapter 1. Where the Wild Genes Are

1 For a portrayal of J. Craig Venter as a pioneering entrepreneurial scientist, see Shapin, *Scientific Life,* 223–226.

2 Website of the National Institutes of Health, http://grants.nih.gov/grants/funding/women_min/guidelines_amended_10_2001.htm, accessed September 11, 2013.

3 Lock and Nguyen, *Anthropology of Biomedicine,* 353.

4 Hacking, "Making Up People."

5 Furnivall, *Netherlands India.*

6 See Nonini, *"Getting By."*

7 See Goh, Gabrielpillai, Holden, and Choo, *Race and Multiculturalism in Malaysia and Singapore.*

8 Chua, "Cost of Membership in Ascribed Community," 174.

9 Mitchell, *Rule of Experts.*

10 Heng and Devan, "State Fatherhood," 204–205; Ong, *Flexible Citizenship,* 68–72.

11 Coincidentally, in 2003, the United Kingdom finally ended its ban on homosexuality. I suspect this had an influence on relaxing Singapore's social attitudes toward modern forms of homosexual and transsexual practices.

12 As well as perhaps beyond what is denoted by "nature," "wild," "culture," "lab," "virtual," "organic," and so on, but only after anthropologists and other human and social scientists vet each of these through contemporary instance work. Also does beyond calculation imply beyond the imaginable too?

13 See Callon and Rabeharisoa, "Research 'in the Wild,'" for a similar use of "in the wild" to indicate the division between already-gathered data versus information to be gathered.

14 Fujimura and Rajagopalan, "Different Differences."

15 Kelty, "Preface."

16 Rabinow, "Artificiality and Enlightenment."

17 Latour, "Visualisation and Cognition," 7–9.

18 The National Cancer Act of 1971 intended to eradicate cancer as a major cause of death.

19 Lim, "Transforming Singapore Health Care," 462.

20 Along with the Biopolis initiative, a Singapore Medicine initiative aims to develop the island into a major medical tourist destination. There is a need to balance this commercialization of health care services (over a million foreign patients a year) and to keep domestic health costs affordable.

21 See Mitchell, Breen, and Enzeroth, "Genomics as Knowledge Enterprise."

22 International Hapmap Project, http://hapmap.ncbi.nlm.nih.gov/, accessed March 22, 2014.

23 Palsson and Rabinow, "Icelandic Controversy," 92.

24 Fortun, *Promising Genomics*, 232.

25 China's drug markets, growing at 25 percent per year compared to 2–5 percent in the West, is the new focus of big drugmakers. S. S. Wang and J. D. Rockoff, "Drug Research Gets New Asian Focus," *Wall Street Journal*, November 15, 2010, B1–B2.

26 Clancey, "Intelligent Island to Biopolis."

27 Cord Blood Bank of Singapore, http://www.scbb.com.sg, accessed May 30, 2012.

28 See Ong, "Lifelines."

29 See Duster, *Backdoor to Eugenics*.

30 By "Orientals," Dr. Wu used the old-fashioned Western term for "Asians," perhaps ironically, with no explicit derogatory connotations.

31 Chew and Tai, *Study Protocol for the Singapore Consortium*, 1.

32 Latour, "Visualisation and Cognition," 13–14.

33 See my earlier formulation of this point in Ong, "Milieu of Mutations."

34 Wil S. Hylton, "Craig Venter's Bugs Might Save the World," *New York Times Magazine*, http://www.nytimes.com/2012/06/03/magazine/craig-venters-bugs-might-save-the-world.html?pagewanted=all&_r=0, accessed May 30, 2012.

35 Quoted in J. Yang, "The Stats behind the Medical Science," *Today* (National University of Singapore), March 31, 2012, p. 22.

Chapter 2. An Atlas of Asian Diseases

1 A strong proponent of this view is Reardon, *Race to the Finish*.

2 See Ong, "Lifelines," 195.

3 See, e.g., Keck, "Hong Kong as a Sentinel Post."

4 Ong, "Introduction."

5 Deister, "Laboratory Life of the Mexican Mestizo."

6 Rabinow, *French DNA*.

7 See Fukuyama, *Trust*, who argues that societies that have or invest in a high degree of social trust will allow enterprises to evolve into professionally managed organizations.

8 Foucault, "Of Other Places."

9 Bourdieu, *Logic of Practice*, 197–203.

10 Waldby, "Biobanking in Singapore."

11 A*STAR Research, *Reports on Ethical, Legal, and Social Issues*.

12 Lim and Ho, "Ethical Position of Singapore on Embryonic Stem Cell Research."

13 Rawls, *Theory of Justice*.

14 Ong, "Lifelines."

15 Anderson, *Imagined Communities*.

16 Ong, "Lifelines."

17 For example, the International Meeting on Clinical and Lab Genomic Standards, May 3–5, 2005, Paris, sponsored by Affymetrix.

18 Ong, "Introduction," 38–42.

19 An error-ridden and extreme case is made by Guo in " 'Gene War of the Century?' "

20 See Wilson, "Medical Tourism in Thailand."

21 For decades, cancer treatment was a one-drug-fits-all proposition, although the drugs were developed using mainly Caucasian models of the disease. For instance, major cancer drugs such as Tamoxifen and Herceptin, both for treating breast cancer, and Gleevec, which targets leukemia, have not been effective for treating forms of cancer among patients of Asian ancestry. After all, the efficacy of such drugs depends on a close match to the genome of the tumor.

22 Mukherjee, *Emperor of All Maladies*, 450.

23 Quoted in Paul Gilfeather, "The Wayne Rooney of Research," *Today* (Singapore magazine), April 26, 2010, B2.

24 A*STAR Research, "On the Trail of an Elusive Tumor."

25 Although most of his collaborations are with colleagues in China, Lin had a joint project on breast cancer with scientists at the Karolinska Institute in Sweden.

26 Haraway, *Modest_Witness@Second_Millennium*, 37 (italics in the original).

Chapter 3. Smoldering Fire

1 See Franklin and McKinnon, *Relative Values*, 11.

2 Ong, "Lifelines."

3 See, e.g., Wailoo, *How Cancer Crosses the Color Line*.

4 Angelina Jolie, "My Medical Choice," *New York Times*, May 14, 2013.

5 Anemona Hartocollis, "Research Partnership Will Seek Human Subjects for Cancer Work," *New York Times*, April 3, 2015, A16–A17.

6 Mukherjee, *Emperor of All Maladies*, 455–457.

7 A leading work on this approach is Mol, *Logic of Care*.

8 Deleuze, *Lectures by Gilles Deleuze*.

9 See Deleuze and Guattari, *Thousand Plateaus*, 441.

10 Berlant, *Cruel Optimism*, 1.

11 Agamben, *Homo Sacer*.

12 Foucault, *Security, Territory, Population*, 20–23.

13 Anthropological analysis of the ethical challenges of work in refugee camps is fundamentally about the politics of sheer survival. See, e.g., Redfield, *Life in Crisis*.

14 For another view of "hope" that depends on projecting a specific temporality of deferral that allows one to survive an intolerable present, see Miyasaki, *Method of Hope*.

15 This is contrary to Foucault's implication that biopolitics seeks to erase death by making it publicly invisible. See his *Society Must Be Defended*, 239–264.

16 See Rabinow, "Artificiality and Enlightenment"; Rose and Novas, "Biological Citizenship"; Nguyen, "Antiretroviral Globalism"; Gibbon and Novas, *Biosocialities*; and van Hoyweghen, "Taming the Wild Life of Genes by Law?"

17 Dumit, *Drugs for Life*.

18 Buchanan, "Deleuze and Geophilosophy."

19 Jasanoff, *Designs on Nature*, 26.

20 Fullwilley, "Biologistical Construction of Race."

21 Quote from Steve West, in "Interview with IPASS Trial and Leading Lung Cancer Researcher Tony Mok," *Global Resource for Advancing Cancer Education*, April 19, 2011, http://www.cancergrace.org/lung/2011/04/19/tony-mok-interview-part-1/, accessed August 16, 2012.

22 "IPASS (IRESSA Pan-Asia Study) was an open label, randomised, parallel-group study that assessed the efficacy, safety and tolerability of IRESSA versus doublet chemotherapy (carboplatin/paclitaxel) as 1st line treatment in a clinically selected population of patients from Asia." See http://www.iressa.com/ipass-study/.

23 Dr. Kwong, "New Cancer Drugs" talk, Biopolis, April 8, 2010.

24 Victoria Colliver, "Racial Diversity Vital to Clinical Trials," *San Francisco Chronicle*, May 21, 2014, D1, D5.

25 Jessica Yeo, "New Pill for Advanced Lung Cancer Patients," *Healthy Me—At Home and Away* (Singapore), December 19, 2011, http://beinghealthyhomeandaway.blogspot.com/2011/12/new-pill-for-advanced-lung-cancer.html, accessed November 11, 2011.

26 Haraway, *Modest_Witness@Second_Millennium*, 70.

27 China's drug markets, growing at 25 percent per year compared to 2–5 percent in the West, is the new focus of big drugmakers. S. S. Wang and J. D. Rockoff, "Drug Research Gets New Asian Focus," *Wall Street Journal*, November 15, 2010, B1–B2.

28 Rose and Novas, "Biological Citizenship."

29 Haraway, *Modest_Witness@Second_Millennium*, 43–44.

30 Rabinow, *French Modern*.

Chapter 4. The Productive Uncertainty of Bioethics

1 Fernando Coronil, "Perspectives on Tierney's *Darkness in El Dorado*," *Current Anthropology* 42, no. 2 (2001): 265–276.

2 Briggs and Mantini-Briggs, *Stories in a Time of Cholera*.

3 See, e.g., Hayden, *When Nature Goes Public*.

4 Sunder Rajan, "Experimental Machinery of Global Clinical Trials," 61–63.

5 Petryna, *When Experiments Travel*, 46.

6 Wallerstein, *Modern World-System, Vol. I*.

7 Petryna, *When Experiments Travel*.

8 Yarborough, "Increasing Enrollment in Drug Trials," 442.

9 Sunder Rajan, *Biocapital*, 60. Fingering Ranbaxy, the FDA has repeatedly banned imports of generic drugs from India that were suspected of being manufactured under "invented safety standards." See "Ranbaxy's Chronic Maladies," *Economist*, September 21, 2013, 65.

10 For a study of the technical requirements of pharmaceutical competition in another emerging country, see Lakoff, *Pharmaceutical Reason*.

11 See Ong and Chen, *Asian Biotech*.

12 See Talbot, *Bioethics*, 16–20.

13 Kaufman, "Construction and Practice of Medical Responsibility"; Rapp, "Moral Pioneers."

14 Hoeyer and Jensen, "Transgressive Ethics."

15 Kowal, "Orphan DNA," 562. Italics in the original.

16 Britain is a leader in formulating bioethical guidelines for human experimentation. The Foundation for Genomics and Population Health states, "The four principles should . . . be thought of as the four moral nucleotides that constitute moral DNA— capable, alone or in combination, of explaining and justifying all the substantive and universalisable moral norms of health care ethics." http://www.phgfoundation .org/tutorials/moral.theories/6.html, accessed December 24, 2013.

17 See, e.g., Cohen, "Operability, Bioavailability, and Exception."

18 DNA sequencing involves reading out the spelling (A, C, G, T bases) of the genetic code. Field technicians collect samples by pricking the fingers of donors and blotting the blood on filter paper. The samples on fragments of paper are then sent to labs for genetic sequencing. DNA can also be collected from saliva.

19 See Keller, *Century of the Gene*.

20 See Rheinberger, *Epistemology of the Concrete*, which argues for the importance of the epistemic object—scientific equipment, models, and concept-objects such as the gene—that is, concrete matter, not just a discursive field, in the epistemology of science.

21 Taussig, "Reification and the Consciousness of the Patient."

22 Geertz, *Local Knowledge*, 73–93.

23 Kowal, "Orphan DNA."

24 See Cooper and Waldby, *Clinical Labor*.

25 Mukherjee, *Emperor of All Maladies*.

26 Dr. Kwong, "New Cancer Drugs" talk, Biopolis, April 8, 2010.

27 Ong, "Scales of Exception."

28 Martin, *Flexible Bodies*.

29 Genentech in California had developed such drugs, such as Herceptin, to treat HER2-positive breast cancer found mainly in Western women.

Chapter 5. Virtue and Expatriate Scientists

1 One such figure with medical training is a former prime minister, Dr. Mohammed Mahathir of Malaysia. Like many political leaders in his cohort, Mahathir imbibed colonial ideas of social Darwinism that drove his racialist vision for modern Malaysia. Before he became prime minister in the 1980s, Mahathir wrote *The Malay*

Dilemma (New York: Times Book International, 1970). He argues that Malays, an easygoing people bred in tropical ease, would always be bested by immigrant Chinese who had evolved in conditions of harsh climate and fierce competition. He suggested there was a genetic basis to this unequal ethnic competition in Malaysia, thus justifying his "affirmative action policy" favoring *majority* Malay (*bumiputra*) rule, which dominates minority ethnic Chinese and Indians. There is an echo of such notions of "soft versus hardy races" in his Singaporean counterpart Lee Kuan Yew's view that in the island-state the presumably more robust ethnic Chinese should dominate (see chapter 1).

2 Alatas, *Myth of the Lazy Native.*
3 See, e.g., Adams, *Doctors for Democracy.*
4 Shapin, *Scientific Life*, 47–91.
5 Weber, "Science as a Vocation."
6 Shapin, *Scientific Life*, 264–267.
7 Brenner, "Genomics and a Scientifically Responsible Ethics."
8 Weber, "Science as a Vocation."
9 MacIntyre, *After Virtue.*
10 Collier and Lakoff, "Regimes of Living," 23–25.
11 See Lakoff and Collier, *Biosecurity Intervention.*
12 Weber, *On Charisma and Institution Building.*
13 Edison Liu landed at the Jackson Labs in Maine. He has also set up a genome center at the University of Connecticut, employing some old Biopolis researchers in his project there.
14 Besides working with partners at Duke University, Johns Hopkins University, and MIT, Biopolis researchers have collaborated with colleagues at University of California, San Diego; University of Illinois at Urbana-Champaign; University of Washington; the Fred Hutchinson Cancer Research Center in the United States; the Karolinska Institute in Sweden; Imperial College, Cambridge University; Dundee University; the French Atomic Energy Commission in Europe; the Australian National University; and RIKEN in Japan.
15 See Ong, "Please Stay."
16 Inland Revenue Authority of Singapore, http://www.iras.gov.sg/irasHome/default .aspx, accessed May 13, 2015.
17 See Ong, "Biocartography."
18 Olds and Thrift, "Cultures on the Brink."

Chapter 6. Perturbing Life

1 Luhmann, *Observations on Modernity*, 71.
2 Haraway, *Simians, Cyborgs, and Women*, 203–231.
3 Luhmann, *Observations on Modernity*, 75–108.
4 Jasanoff, *Designs on Nature*, chapter 7.
5 Quotation from a presentation at the Singapore-Australia Joint Symposium on Stem Cells and Bioimaging, Biopolis, Singapore, May 24–25, 2010.
6 Quotation from a presentation at the symposium "Stem Cells Applications: Reprogramming through Therapeutics," Biopolis, Singapore, April 22, 2010.

7 Quoted in E. Brown, "Reverting Cells to Their Embryonic State, without the Embryos," *Los Angeles Times*, November 27, 2010.

8 A*STAR Research, *Reports on Ethical, Legal, and Social Issues in Human Stem Cell Research*; National Survey of R&D in Singapore 2003.

9 Franklin, "Embryonic Economies."

10 Cooper and Waldby, *Clinical Labor*, 115. See also Thompson, *Making Parents*.

11 Ong, "Lifelines," 208.

12 Quote from Robert Williamson, at the Singapore-Australia Symposium on Stem Cells and Bioimaging.

13 Quoted in "Kyoto Prize to UCSF SC Researcher," *San Francisco Chronicle*, June 18, 2010, C2.

14 Quoted during an interview at the Singapore-Australia Joint Symposium on Stem Cells and Bioimaging.

15 Belmont, Ellis, Hochedlinger, and Yamanaka, "Induced Pluripotent Stem Cells and Reprogramming."

16 Quoted from an interview during the symposium "Stem Cells Applications: Reprogramming through Therapeutics."

17 Thompson, "Asian Regeneration?," 105–108.

18 Sleeboom-Faulkner, "Embryo Controversies," 230–231.

19 Chen, "Feeding the Nation."

20 Liu, "Making Taiwanese (Stem Cells)."

21 Ong, "Lifelines," 196.

22 Colman, "Stem Cell Research in Singapore."

23 Franklin, "Embryonic Economies," 86.

24 Thompson, "Asian Regeneration?," 105.

25 Ong, *Spirits of Resistance and Capitalist Discipline*, 152.

26 Zhou et al., "Generation of Induced Pluripotent Stem Cells Using Recombinant Proteins."

27 Skloot, *The Immortal Life of Henrietta Lacks*, comes to mind. The book is an extended account of the casual appropriation of Henrietta Lacks's cancerous cells decades ago and their subsequent distribution as research materials to labs all over the world. The publicity surrounding the Lacks case focuses on the issue of informed consent in the unauthorized uses of Lacks's cells and the billions of U.S. dollars in profit generated thereby for research institutions but not for Lacks's surviving family.

28 Neyrat, "Birth of Immunopolitic."

Chapter 7. A Single Wave

Epigraph is quoted in Seema Singh, "Early Human Settlers Used Southern Coastal India to Enter Asia," *Livemint*, December 10, 2009, http://www.livemint.com/2009/12/10225237/early-human-settlers-used-sout.html, accessed February 15, 2013.

1 Cyranoski, "Asia Populated in One Migratory Swoop"; Singh, "Early Human Settlers."

2 Cavalli-Sforza, "Human Genome Diversity Project."

3 Lakoff, "Generic Biothreat," 401.

4 Haraway, *Primate Visions*, 5.

5 Reardon, *Race to the Finish*.

6 Marks, "Commentary," 749.

7 See Dawkins, *River out of Eden*.

8 Dawkins, *Selfish Gene*.

9 Ong, "Scales of Exception."

10 Fischer, "Biopolis."

11 Duara, "Discourse of Civilization and Pan-Asianism." It is important to highlight the fact that Japanese ultra-nationalists sponsored the propaganda ideas of the "Greater East Asia Co-prosperity Sphere" and the "manifest destiny" of Japan to be the leader of all Asia against Western imperialist powers. But invaded Asian nations (Manchuria, Mongolia, China, all of Southeast Asia) soon experienced Japan as just another kind of oppressive imperialist power and took up arms to liberate their own peoples. Japanese military violence under this "new order" along with the phenomenon of enslaved "comfort women" are political reasons for why Japan has not been able to exert leadership in the region since the end of the war in 1945.

12 Lewis and Wigen, *Myth of Continents*.

13 The Hapmap was succeeded by a more inclusive project. In 2012, the 1000 Genomes project, describing the genomes of fourteen groups across Europe, the Americas, Africa, and Asia, published its findings. In this new catalog of genetic variation, the genomes comprised individuals who are defined by a mix of cultural, biological, and geographic elements: racial ancestry, ethnicity, nationality, and location. See 1000 Genomes Project Consortium, "Integrated Map of Genetic Variation from 1,092 Human Genomes," 11632.

14 The Pan-Asian SNP database sampled 1,719 unrelated individuals among seventy-one populations from China, India, Indonesia, Japan, Malaysia, the Philippines, Singapore, South Korea, Taiwan, and Thailand.

15 Jorde and Wooding, "Genetic Variation, Classification and 'Race.'"

16 HUGO Pan-Asian SNP Consortium, "Mapping Human Genetic Diversity in Asia," 1541.

17 Edison Liu, "Genetics of Race in Clinical Trials," talk presented at the National Cancer Institute conference, "The Meaning of Race in Science," New York, April 9, 1997.

18 HUGO Pan-Asian SNP Consortium, "Mapping Human Genetic Diversity in Asia," 1542–1544.

19 In the 1980s, an Italian geneticist gathered common known variations in simple genes (classical polymorphisms) and traced five genetic maps of earlier migrations associated with modern-day Greeks, Basques, and Finns, among other groups. See Cavalli-Sforza, "Human Genome Diversity Project."

20 Cyranoski, "Asia Populated in One Migratory Swoop."

21 Vikrant Kumar, "Scientific Consortium Maps the Range of Genetic Diversity in Asia, and Traces the Genetic Origins of Asian Populations," Genome Institute of Singapore, press release, December 11, 2009.

22 Raja Murthy, "India Is 'Thailand' to Asia, Say Scientists," *Asia Times Online*, December 19, 2009, http://www.atimes.com/atimes/South_Asia/KL19Df03.html, accessed February 15, 2013.

23 Quoted in Kumar, "Scientific Consortium Maps the Range of Genetic Diversity in Asia."

24 Normille, "SNPS Study Supports Southern Migration Route to Asia," 1470.

25 Quoted in Kumar, "Scientific Consortium Maps the Range of Genetic Diversity in Asia."

26 Quoted in Kumar, "Scientific Consortium Maps the Range of Genetic Diversity in Asia."

27 *Wall Street Journal* (Asia), "Singapore Likes a Crowd," January 29, 2013.

28 Cho, Do, Chandrasekaran, and Kan, "Identifying Research Facilitators in an Emerging Asian Research Area."

29 Shapin, *Scientific Life*.

Chapter 8. "Viruses Don't Carry Passports"

1 George, *Singapore*; Clancey, "Intelligent Island to Biopolis."

2 See Lakoff, "Generic Biothreat."

3 See the issue on figures of warning, *Limn* no. 3 (2013), http://limn.it/preface-sentinel-devices-2/.

4 Latour, "Visualisation and Cognition," 17–19.

5 Barry, *Great Influenza*; Garrett, *Coming Plague*; Preston, *Hot Zone*; Wolfe, *Viral Storm*; Quammen, *Spillover*.

6 See Garrett, *Coming Plague*.

7 See, e.g., Haraway, *When Species Meet (Posthumanities)*; Kirksey and Helmreich, "Emergence of Multispecies Ethnography"; Lowe, "Viral Clouds"; and Brown and Kelly, "Material Proximities and Hotspots."

8 See, e.g., Whitington, "Fingerprinting, Bellwether, Model Event"; and Choy and Zee, "Condition—Suspension."

9 Collier and Lakoff, "Distributed Preparedness."

10 Keck and Lakoff, "Preface: Figures of Warning."

11 Keck, "Hong Kong as a Sentinel Post."

12 Whitington, "Fingerprinting, Bellwether, Model Event."

13 Brown and Kelly, "Material Proximities and Hotspots," 280; see also Lowe, "Viral Clouds."

14 Foucault, *Madness and Civilization*. He traces logics of confinement and segregation for managing the fear of diseases back to the end of the eighteenth century, when the medical profession introduced the practice of removing "the mad" from society at large. In *The Birth of the Clinic*, Foucault argues that "tertiary spatialization" associated with medical observations include "all the gestures by which, in a given society, a disease is circumscribed, medically invested, isolated, divided up into closed, privileged regions, or distributed throughout cure centers" (16).

15 Law and Mol, "Situating Technoscience."

16 Foucault, *Security, Territory, Population*, 45.

17 Quammen, *Spillover*, 512.

18 Ong, "Zoning Technologies in East Asia."

19 Ong, "Scales of Exception"; Ong, "Milieu of Mutations."

20 See Fearnley, "Birds of Poyang Lake."

21 See Quammen, *Spillover*, 187–192.

22 See Ong, "Zoning Technologies in East Asia."

23 A*STAR Research, "Strengthening the Fight against Dengue Fever." Here was an echo of the claims about Southern Chinese genetic variability and their special vulnerability to respiratory and gastrointestinal infections linked to coronaviruses, including SARS.

24 National Environment Agency, "Operations Strategy," http://www.dengue.gov.sg /subject.asp?id=34, accessed March 27, 2015.

25 Duke-NUS Graduate Medical School, "Emerging Infectious Diseases," http://www .duke-nus.edu.sg/research/signature-research-programs/emerging-infectious -diseases, accessed June 21, 2013.

26 A*STAR Research, "Strengthening the Fight against Dengue Fever."

27 Pharmabiz.com, "SIgN Discovers Breakthrough Strategy to Disarm Dengue Virus; Brings Hope for an Effective Vaccine," August 14, 2013, http://www.a-star.edu.sg /Portals/32/Files/PharmaBiz.pdf, accessed July 14, 2015.

28 Quammen, *Spillover*, 512.

29 Chua et al., "Previously Unknown Reovirus of Bat Origin."

30 For a Singapore vs. China comparison of the "human security and public health" reactions to SARS, see Curley and Thomas, "Human Security and Public Health in Southeast Asia."

31 Ong, "Assembling around SARS."

32 Cheong, "SARS of 'The Plague.'"

33 Professor Tan Chorh Chuan, the president of NUS, cited in "Duke-NUS Spurs Research into Emerging Infectious Diseases with Major Meeting of Global Experts," Duke-NUS media release, December 8, 2009.

34 Centers for Disease Control and Prevention, "Severe Acute Respiratory Syndrome—Singapore," MMWR, May 9, 2003, http://www.cdc.gov/mmwr/preview /mmwrhtml/mm5218a1.htm, accessed July 14, 2015.

35 Centers for Disease Control and Prevention, "Remembering SARS: A Deadly Puzzle and Efforts to Solve It," MMWR, April 11, 2013, http://www.cdc.gov/about /history/sars/feature.htm, accessed July 14, 2015.

36 Centers for Disease Control and Prevention, "Remembering SARS."

37 Singapore Ministry of Health, "Regional Emerging Disease Initiative (REDI) Centre."

38 Wolfe, *Viral Storm*, 179–181.

39 Globally, the USAID agency has an emerging pandemic threat initiative that is building a database on viruses and sending fieldworkers to train hunters on how to handle game without spreading the risk of infection.

40 See Nguyen, *Republic of Therapy*, 180.

Chapter 9. The "Athlete Gene" in China's Future

1 Henry Yang Huangming, at the Asia Vision 21 conference, National University of Singapore, April 30, 2010.

2 See Rabinow and Bennett, *Designing Human Practices*.

3 Henry Sender, "Chinese Innovation: BGI's Code for Success," *Financial Times*, February 16, 2015.

4 Shu-Ching Jean Chen, "Genomic Dreams Coming True in China," *Forbes Asia*, August 28, 2013, http://www.forbes.com/sites/forbesasia/2013/08/28/genomic -dreams-coming-true-in-china/, accessed April 6, 2015.

5 Chen, "Genomic Dreams Coming True in China."

6 Cyranoski, "Sequence Factory."

7 Specter, "Gene Factory."

8 Larson, "Inside China's Genome Factory."

9 Email from the administrative office, BGI Genomics, Shenzhen, May 5, 2012.

10 Bowker, "Time, Money, Biodiversity." Bowker identifies two possible modalities for incorporating diverse objects into a regime of truth about the way the world is. Ideally, a "modality of implosion," he suggests, collapses representations of several registers into a single form of planetary ecological sustainability. But, he argues, the dominant modality of particularity tries to list every last living thing and give it a particular calculable value according to "our emergent late-capitalist ethnos" (107–108, 119–120). I consider this binary framing too stark, dividing sustainability from capitalism. In any case, BGI as a research-oriented private company hopes to offer scientific solutions that can promote the ideal of planetary sustainability.

11 Yang, quoted at the Asia Vision 21 Conference.

12 BGI Americas, "Accelerating Scientific Breakthroughs," http://bgiamericas.com /scientific-expertise/collaborative-projects/, accessed May 24, 2012.

13 Wang Jian stepped down as president in 2015.

14 Included as well are five food crops that a legendary emperor proclaimed were sacred: soybeans, rice, wheat, barley, and millet.

15 Eli Lilly Corporation, "Lilly, Merck, and Pfizer Join Forces to Accelerate Research and Improve Treatment of Lung and Gastric Cancers in Asia," https://investor.lilly .com/releasedetail.cfm?ReleaseID=446317, accessed May 5, 2015.

16 Mullaney, *Coming to Terms with the Nation.*

17 X. Cai and C. Wang, "Immortal Genome—Preserving China's Ethnic Bloodline," *China Pictorial*, 2006, http://www.rmhb.com.cn/chpic/htdocs/english/200604/5 -2.htm, accessed March 5, 2014.

18 Gan et al., "Pinghua Population."

19 BGI Shenzhen, "Yuanhuang—the First Asian Diploid Genome," http://yh .genomics.org.cn/index.jsp, accessed May 24, 2012.

20 Harrell, *Cultural Encounters on China's Ethnic Frontiers*; Litzinger, *Other Chinas*; Schein, *Minority Rules*; Gladney, *Dislocating China.*

21 Schein, *Minority Rules.*

22 Vasantkumar, "Han at Minzu's Edges."

23 Gladney, *Dislocating China.*

24 Anderson, *Imagined Communities.*

25 Sung, "Chinese DNA," 264.

26 Adams, Erwin, and Le, "Governing through Blood," 167–168.

27 Erin Podolak, "Fastest Case of Adaptation Documented in Tibetans," *Biotechniques*, July 19, 2010, http://www.biotechniques.com/news/fastest-case-of-adaptation -documented-in-Tibetans/biotechniques-299799.html, accessed March 7, 2014.

28 Nielsen is not disguised here because his role in the Tibetan DNA project has been widely publicized.

29 E. Huerta-Sánchez, X. Jin, Z. B. Asan, B. M. Peter, N. Vinckenbosch, et al., "Altitude Adaptation in Tibetans Caused by Introgression of Denisovan-like DNA," *Nature*, 2014, doi:10.1038/nature13408.

30 Quoted in Elizabeth Weise, "Rapid Evolution Seen in Tibetans," *USA Today*, July 2, 2010, http://usatoday30.usatoday.com/news/world/2010-07-01-Tibet_N.htm, accessed March 7, 2014.

31 Žižek, "Philosophy, the 'Known Unknowns.'"

32 For an appeal to distinguish between "arboreal" and "rhizomatic" modalities in the human sciences, see Deleuze and Guattari, *Thousand Plateaus*, chapter 1.

33 Email from the administrative office, BGI Genomics, Shenzhen, May 5, 2012.

Epilogue

1 The triple-tower complex was designed by Moshe Safdie to capture the open geometry of the city-state.

2 Singapore Tourism Board, "Helix Bridge," http://www.yoursingapore.com/content/traveller/en/browse/see-and-do/arts-and-entertainment/architecture/helix-bridge.html, accessed July 9, 2015.

3 Deleuze and Guattari, *Thousand Plateaus*.

4 Ong Keng Sen, "Director's Note," April 2014, http://spoletousa.org/wp-content/uploads/2013/11/Facing-Goya.pdf, accessed April 29, 2015.

5 For important anthropological studies, see Kleinman, *Patients and Healers*; Unschuld, *Medicine in China*; Hsu, *Transmission of Chinese Medicine*; Chen, *Breathing Spaces*; Zhan, *Other-Worldly*; and Farquhar and Zhang, *Ten Thousand Things*.

6 Kleinman, *Healing Narratives*.

7 Larson, "Inside China's Genome Factory," 4–5.

8 See Anagnost, "Corporeal Politics of Quality (Suzhi)."

9 Email received from the BGI Shenzhen administrative unit, May 5, 2012.

10 Greenhalgh, *Just One Child*.

11 There is disagreement as to the measure of the actual length of "Chinese civilization." Recorded history of continuously "Chinese" culture is estimated at 3,700 years.

12 Greenhalgh, *Just One Child*.

13 Quoted in Aleks Eror, "China Is Engineering Genius Babies," *Vice*, March 15, 2013, http://www.vice.com/read/chinas-taking-over-the-world-with-a-massive-genetic-engineering-program, accessed May 13, 2015.

14 The National Academy of Medicine, the Royal Society (United Kingdom), and the Chinese Academy of Sciences have come together to establish ethical guidelines for using the gene-editing technique. For an American perspective that registers unease about Chinese researchers using the Crispr method, see Gina Kolata, "Chinese Scientists Edit Genes of Embryos, Raising Concerns," *New York Times*, April 23, 2015, http://www.nytimes.com/2015/04/24/health/chinese-scientists-edit-genes-of-human-embryos-raising-concerns.html?_r=0, accessed April 23, 2015.

15 Žižek, "Philosophy, the 'Known Unknowns,'" 137, 139.

16 Neyrat, "Birth of Immunopolitic."

17 See Martin Enserink, "Golden Rice Not So Golden for Tufts," *Science*, September 18, 2013, http://news.sciencemag.org/asiapacific/2013/09/golden-rice-not-so-golden-tufts, accessed April 4, 2015.

18 Foucault, *Order of Things*.

This bibliography does not include newspaper articles and most ephemeral institutional reports and business publications. Those sources are cited in full within the endnotes.

Singapore: Official Reports and News Articles

A*STAR Research. "On the Trail of an Elusive Tumor." Published online September 1, 2010.

————. *Reports on Ethical, Legal, and Social Issues in Human Stem Cell Research, Reproductive and Therapeutic Cloning.* National Survey of R&D in Singapore, 2002.

————. "Singapore's Biopolis: A Success Story." Media press release. http://w.w.w.a-star.edu.sg/Media/News/Press-Releases/ID/1893/Singapore-Biopolis-A-Success-Story.aspx. Accessed February 10, 2015.

————. "Strengthening the Fight against Dengue Fever." Published online September 25, 2013. https://www.research.a-star.edu.sg/feature-and-innovation/6793/strengthening-the-fight-against-dengue-fever. Accessed March 23, 2016.

Cheong, P. K. "SARS of 'The Plague.'" *SMA News* (Singapore Medical Association) 35, no. 6 (June 2003): 23–26.

Chew, S. K., and E. S. Tai. *Study Protocol for the Singapore Consortium for Cohort Studies (SCCS).* September 6, 2007. Unpublished manuscript.

Lim, Sylvia, and Calvin Ho. "The Ethical Position of Singapore on Embryonic Stem Cell Research." *SMA News* (Singapore Medical Association) 35, no. 6 (2003): 21–23.

Singapore Ministry of Health. "Regional Emerging Disease Initiative (REDI) Centre." 2011. http://www.moh.gov.sg/content/moh_web/home/Publications/information_papers/2003/regional_emergingdiseaseinterventionredicentre.html. Accessed November 14, 2013.

Tan, W. "Warning against Complacency, 10 Years after SARS." *Today* (National University of Singapore), March 19, 2013.

Yang, J. "The Stats behind the Medical Science." *Today* (National University of Singapore), March 31, 2012.

Secondary Sources

Adams, Vincanne. *Doctors for Democracy: Health Professionals in the Nepal Revolution.* Cambridge, UK: Cambridge University Press, 1998.

Adams, Vincanne, Kathleen Erwin, and Phouc V. Le. "Governing through Blood: Biology, Donation, and Exchange in Urban China." In *Asian Biotech: Ethics and Communities of Fate,* edited by Aihwa Ong and Nancy N. Chen, 167–189. Durham, NC: Duke University Press, 2010.

Agamben, Giorgio. *Homo Sacer: Sovereign Power and Bare Life.* Stanford, CA: Stanford University Press, 1998.

Alatas, Syed Hussein. *The Myth of the Lazy Native: A Study of the Image of the Malays, Filipinos and Javanese from the 16th to the 20th Century and Its Function in the Ideology of Colonial Capitalism.* London: Routledge, 2010.

Anagnost, Ann. "The Corporeal Politics of Quality (Suzhi)." *Public Culture* 16, no. 2 (Spring 2004): 189–208.

Anderson, Benedict. *Imagined Communities: Reflections on the Origins and Spread of Nationalism.* London: Verso, 2006.

Aristotle. *Politics/Aristotle.* Book 5. Translated by Benjamin Jowitt. Mineola, NY: Dover, 2000.

Barry, Andrew. "Ethical Capitalism." In *Global Governmentality,* edited by W. Larner and W. Walters. New York: Routledge, 2004.

Barry, John M. *The Great Influenza: The Epic Story of the Deadliest Plague in History.* New York: Penguin, 2005.

Bateson, Gregory. *Steps to an Ecology of Mind: Collected Essays in Anthropology, Psychiatry, Evolution and Epistemology.* Chicago: University of Chicago Press, 1999. First published 1972.

Beck, Ulrich. *Risk Society: Towards a New Modernity.* London: Sage, 1992.

Bell, Daniel. *Communitarianism and Its Critics.* New York: Oxford University Press, 1993.

Belmont, J. C. I., J. Ellis, K. Hochedlinger, and S. Yamanaka. "Induced Pluripotent Stem Cells and Reprogramming: Seeing the Science through the Hype." *Nature Reviews Genetics,* published online October 27, 2009.

Benjamin, Walter. *The Work of Art in the Age of Its Technological Reproducibility, and Other Writings on Media.* Cambridge, MA: Belknap Press, 2008. First published 1936.

Berlant, Lauren. *Cruel Optimism.* Durham, NC: Duke University Press, 2011.

BGI China. "1,000 Plants and Animals Reference Genomes Project." 2011. http://ldl .genomics.cn/page/pa-research.jsp. Accessed June 6, 2014.

Bourdieu, Pierre. *The Logic of Practice.* Stanford, CA: Stanford University Press, 1992.

Bowker, Geoffrey. "Biodiversity Datadiversity." *Social Studies of Science* 30, no. 5 (2001): 643.

———. "Time, Money, Biodiversity." In *Global Assemblage: Technology, Politics and Ethics as Anthropological Problems,* edited by Aihwa Ong and Stephen J. Collier, 107–123. Malden, MA: Wiley-Blackwell, 2005.

Brenner, Sydney. "Genomics and a Scientifically Responsible Ethics." *Biosocieties* 1, no. 1 (March 2006): 7–12.

Briggs, Charles L., and Clara Mantini-Briggs. *Stories in a Time of Cholera: Racial Profiling during a Medical Nightmare*. Berkeley: University of California Press, 2004.

Brown, Hannah, and Ann H. Kelly. "Material Proximities and Hotspots: Toward an Anthropology of Viral Hemorrhagic Fevers." *Medical Anthropology Quarterly* 28, no. 2 (2014): 280–303.

Buchanan, Brian. "Deleuze and Geophilosophy." *Journal for Cultural and Religious Theory* 7, no. 1 (2005): 131–134.

Callon, Michel. "What Does It Mean to Say That Economics Is Performative?" In *Do Economists Make Markets? The Performativity of Economics*, edited by D. Mackenzie, F. Muniesa, and L. Siu. Princeton: Princeton University Press, 2007.

Callon, Michel, and V. Rabeharisoa. "Research 'in the Wild' and the Shaping of New Social Identities." *Technology in Society* 25 (2003): 193–204.

Canguilhem, Georges. *Knowledge of Life,* New York: Fordham University Press, 2008.

———. *The Normal and the Pathological*. Translated by Carolyn R. Fawcett, New York: Zone, 1989.

Cavalli-Sforza, Luca L. "The DNA Revolution in Population Genetics." *Trends in Genetics* 14, no. 2 (February 1998): 60–65.

———. "The Human Genome Diversity Project: Past, Present, and Future." *Nature Reviews Genetics* 6 (2005): 333–340.

Chen, Nancy N. *Breathing Spaces: Qigong, Psychiatry, and Healing in China*. New York: Columbia University Press, 2003.

———. "Feeding the Nation: Chinese Biotechnology and Genetically Modified Foods." In *Asian Biotech: Ethics and Communities of Fate*, edited by Aihwa Ong and Nancy N. Chen, 81–92. Durham, NC: Duke University Press, 2010.

Cho, Philip S., H. H. N. Do, M. K. Chandrasekaran, and M.-Y. Kan. "Identifying Research Facilitators in an Emerging Asian Research Area." *Scientometrics* (2013). doi 10.1007/s11192-013-1051-3.

Choy, Tim, and Jerry Zee. "Condition—Suspension." *Cultural Anthropology* 30, no. 2 (2015): 210–223.

Chua, Beng Huat. "The Cost of Membership in Ascribed Community." In *Multiculturalism in Asia*, edited by W. Kymlicka and He Baogang, 170–195. Oxford: Oxford University Press, 2005.

———. "Singapore as Model: Planning Innovations, Knowledge Experts." In *Worlding Cities: Asian Experiments and the Art of Being Global*, edited by Ananya Roy and Aihwa Ong, 29–54. Oxford: John Wiley and Sons, 2011.

Chua, K. B., C. Crameri, A. Hyatt, M. Yu, M. R. Tompang, J. Rosli, J. McEachern, S. Crameri, V. Kumarasamy, B. T. Eaton, and L.-F. Wang. "A Previously Unknown Reovirus of Bat Origin Is Associated with an Acute Respiratory Disease in Humans." *Proceedings of the National Academy of Sciences USA* 104 (2007): 11424–11429.

Clancey, Gregory. "Intelligent Island to Biopolis: Smart Minds, Sick Bodies and Millennial Turns in Singapore." *Science Technology and Society* 17, no. 1 (2012): 13–35.

Clarke, Adele E., Janet K. Shim, Laura Mamo, Jennifer Ruth Fosket, and Jennifer R. Fishman. "Biomedicalization: Technoscientific Transformations of Health, Illness, and U.S. Biomedicine." *American Sociological Review* 68, no. 2 (2003): 161–194.

Cohen, Lawrence. 1999. *No Aging in India: Alzheimer's, the Bad Family, and Other Modern Things*. Berkeley: University of California Press, 2000.

———. "Operability, Bioavailability, and Exception." In *Global Assemblage: Technology, Politics and Ethics as Anthropological Problems*, edited by Aihwa Ong and Stephen J. Collier, 79–90. Malden, MA: Blackwell, 2005.

———. "Where It Hurts: Indian Material for an Ethics of Organ Transplantation." *Daedelus* 128, no. 4 (1999): 135–165.

Collier, Stephen J. "Topologies of Power: Foucault's Analysis of Political Government beyond Governmentality." *Theory, Culture and Society* 26, no. 6 (2009): 78–108.

———. "Vital Systems Security." New York: New School Working Papers, 2009.

Collier, Stephen J., and Andrew Lakoff. "Distributed Preparedness: Notes on the Genealogy of Homeland Security." *Environment and Planning D: Space and Society* 26, no. 1 (2008): 7–28.

———. "Regimes of Living." In *Global Assemblage: Technology, Politics and Ethics as Anthropological Problems*, edited by Aihwa Ong and Stephen J. Collier, 22–39. Malden, MA: Blackwell, 2005.

Collier, Stephen J., and Aihwa Ong. "Global Assemblages, Anthropological Problems." In *Global Assemblage: Technology, Politics and Ethics as Anthropological Problems*, edited by Aihwa Ong and Stephen J. Collier, 3–21. Malden, MA: Blackwell, 2005.

Colman, Alan. "Stem Cell Research in Singapore." *Cell* 132, no. 4 (February 22, 2008): 519–521.

Cooper, Melinda, and Catherine Waldby. *Clinical Labor: Tissue Donors and Research Subjects in the Global Bioeconomy*. Durham, NC: Duke University Press, 2014.

Curley, M., and N. Thomas. "Human Security and Public Health in Southeast Asia: The SARS Outbreak." *Australian Journal of International Affairs* 58, no. 1 (2004): 17–32.

Cyranoski, David. "Asia Populated in One Migratory Swoop." *Scientific American*, December 10, 2009.

———. "The Sequence Factory." *Nature* 464 (March 2010): 22–24.

Dawkins, Richard. *River out of Eden: A Darwinian View of Life*. New York: Basic Books, 1996.

———. *The Selfish Gene*. 2nd edition. Oxford: Oxford University Press, 1990.

Deister, Vivette G. "Laboratory Life of the Mexican Mestizo." In *Mestizo Genomics: Race Mixture, Nation, and Science in Latin America*, edited by P. Wade, C. L. Beltran, E. Restrepo, and R. V. Santos, 161–182. Durham, NC: Duke University Press, 2014.

Deleuze, Gilles. *Difference and Repetition*. Translated and with a preface by Paul Patton. New York: Columbia University Press, 1994. Original published 1969.

———. *The Fold: Liebniz and the Baroque*. Translated and with a foreword by Tom Conley. Minneapolis: University of Minnesota Press, 1992.

———. *Foucault*. Translated by Sean Hand. Foreword by Paul Bové. Minneapolis: University of Minnesota Press, 1988.

———. *Lectures by Gilles Deleuze: Transcripts on Spinoza's Concept of Affect, 2010*. http://www.webdeleuze.com/php/sommaire.htm. Accessed May 11, 2010.

Deleuze, Gilles, and Felix Guattari. *A Thousand Plateaus: Capitalism and Schizophrenia*. Translated by Brian Massumi. Minneapolis: University of Minnesota Press, 1987.

Desrosieres, Alan. *The Politics of Large Numbers: A History of Statistical Reasoning,* Cambridge, MA: Harvard University Press, 1998.

Duara, Prasanjit. "The Discourse of Civilization and Pan-Asianism." *Journal of World History* 12 (2001): 99–130.

Dumit, Joseph. *Drugs for Life: How Pharmaceutical Companies Define Our Health.* Durham, NC: Duke University Press, 2012.

Duster, Troy. *Backdoor to Eugenics.* 2nd edition. New York: Routledge, 2003.

Farquhar, Judith, and Qicheng Zhang, *Ten Thousand Things: Nurturing Life in Contemporary Beijing.* Cambridge, MA: MIT Press, 2012.

Fearnley, Lyle. "The Birds of Poyang Lake: Sentinels at the Interface of Wild and Domestic." *Limn* no. 3 (2013). http://limn.it/the-birds-of-poyang-lake-sentinels-at-the-interface-of-wild-and-domestic/.

Fischer, Michael M. J. *Anthropological Futures.* Durham, NC: Duke University Press, 2009.

———. "Biopolis: Asian Science in the Global Circuitry." *Science, Technology and Society* 18, no. 3 (November 2013): 379–404.

Fortun, Michael. *Promising Genomics. Iceland and deCODE Genetics in a World of Speculation.* Berkeley: University of California Press, 2008.

Foucault, Michel. *The Birth of the Clinic: An Archeology of Medical Perception.* Translated and with a note by A. M. Sheridan Smith. New York: Vintage, 1994. Original published 1963.

———. *Discipline and Punish.* Translated by Alan Sheridan. New York: Vintage, 1975.

———. *The History of Sexuality, Vol. 1, An Introduction.* Translated by Robert Hurley. New York: Vintage, 1978.

———. *Knowledge/Power: Selected Interviews and Other Writings, 1972–77.* Edited by Colin Gordon. Translated by Colin Gordon, Leo Marshall, John Mepham, and Kate Soper. New York: Vintage, 1980.

———. *Madness and Civilization: A History of Madness in the Age of Reason.* Translated by Richard Howard. New York: Vintage, 1988. Original published 1961.

———. "Of Other Places: Utopias and Heterotopias." "Des Espace Autres," March 1967. Translated from the French by Jay Miskowiec, 1984. http://web.mit.edu/allanmc/www/foucault1.pdf.

———. *The Order of Things: An Archeology of the Human Sciences.* Translated by Alan Sheridan. Reissued edition. New York: Vintage, 1994. Original published 1970.

———. *Security, Territory, Population. Lectures at the Collège de France, 1977–1978.* Edited by M. Senellart. Translated by G. Burchell. New York: Palgrave Macmillan, 2007. Original published 1988.

———. *Society Must Be Defended. Lectures at the Collège de France, 1975–1976.* Translated by David Macey. New York: Picador 2003.

Fox Keller, Evelyn. *The Century of the Gene.* Cambridge, MA: Harvard University Press, 2002.

Franklin, Sarah. "Embryonic Economies: The Double Reproductive Value of Stem Cells." *Biosocieties* 1 (2006): 71–90.

Franklin, Sarah, and Susan McKinnon, eds. *Relative Values: Reconfiguring Kinship Studies.* Durham, NC: Duke University Press, 2002.

Fujimura, Joan H., and R. Rajagopalan. "Different Differences: The Use of 'Genetic Ancestry' versus Race in Biomedical Human Genetic Research." *Social Studies of Science* 41 (2011): 5–30.

Fukuyama, Francis. *Trust: Social Virtue and the Creation of Prosperity*. New York: Free Press, 1996.

Fullwilley, Duana. "The Biologistical Construction of Race: 'Admixture' Technology and the New Genetics Medicine." *Social Studies of Science* 38 (October 2008): 695–735.

———. "The Molecularization of Race: Institutionalizing Human Differences in Pharmacogenetics Practice." *Science as Culture* 16, no. 1 (2007): 1–30.

Furnivall, J. S. *Netherlands India: A Study of Plural Economy*. Cambridge, UK: Cambridge University Press, 1939.

Gan, R.-J., S.-L. Pan, L. F. Mustavich, Z.-D. Qin, X.-Y. Cai, J. Qian, C.-W. Liu, J.-H. Peng, S.-L. Li, J.-S. Xu, L. Jin, and H. Li. "Pinghua Population as an Exception of Han Chinese's Coherent Genetic Structure." *Journal of Human Genetics* 53 (2008): 303–313.

Garrett, Laurie. *The Coming Plague: Newly Emerging Diseases in a World out of Balance*. New York: Penguin, 1994.

Geertz, Clifford. *Local Knowledge*. New York: Basic Books, 1983.

George, Charian. *Singapore: The Air-conditioned Nation: Essays on the Politics of Comfort and Control, 1990–2000*. Singapore: Landmark, 2000.

Gibbon, Sahra, and Carlos Novas, eds. *Biosocialities, Genetics and the Social Sciences*. London: Routledge, 2009.

Gladney, Dru. *Dislocating China: Muslims, Minorities, and Other Subaltern Subjects*. Chicago: University of Chicago Press, 2004.

Goh, D. P., M. Gabrielpillai, P. Holden, and G. C. Choo, eds. *Race and Multiculturalism in Malaysia and Singapore*. London: Routledge, 2009.

Graeber, David. *Debt: The First 5,000 Years*. Brooklyn: Melville House, 2012.

Greenhalgh, Susan. *Just One Child: Science and Policy in Deng's China*. Berkeley: University of California Press, 2008.

Guo, S-W. "'Gene War of the Century?'" *Science*, December 5, 1997, 1693–1697. http://www.sciencemag.org/cgi/content/full/278/5344/1693a.

Guyer, Jane I. *Marginal Gains: Monetary Transactions in Atlantic Africa*. Chicago: University of Chicago Press, 2004.

———. "Prophesy and the Near Future: Macroeconomic, Evangelical, and Punctuated Time." *American Ethnologist* 34, no. 3 (2007): 409–421.

Hacking, Ian. "Making Up People." *London Review of Books* 28, no. 16 (August 2006): 13–26.

———. *The Taming of Chance (Ideas in Context)*. Cambridge, UK: Cambridge University Press, 1990.

Haraway, Donna. *Modest_Witness@Second_Millennium: FemaleMale_Meets_OncoMouse*. New York: Routledge, 1997.

———. *Primate Visions: Gender, Race, and Nature in the World of Modern Science*. London: Routledge, 1989.

———. *Simians, Cyborgs, and Women: The Reinvention of Nature*. New York: Routledge, 1991.

———. *When Species Meet (Posthumanities)*. Minneapolis: University of Minnesota Press, 2007.

Harrell, Stevan. *Cultural Encounters on China's Ethnic Frontiers*. Seattle: University of Washington Press, 1996.

Hartigan, John. "Mexican Genomics and the Roots of Racial Thinking." *Cultural Anthropology* 28 (2013): 372–395.

Hayden, Cori. *When Nature Goes Public: The Making and Unmaking of Bioprospecting in Mexico*. Princeton: Princeton University Press, 2003.

Helmreich, Stefan. *Alien Ocean: Anthropological Voyages in Microbial Seas*. Berkeley: University of California Press, 2009.

Heng, Geraldine, and Janadas Devan. "State Fatherhood: The Politics of Nationalism, Sexuality and Race in Singapore." In *Bewitching Women, Pious Men*, edited by Aihwa Ong and Michael Peletz, 195–215. Berkeley: University of California Press, 1995.

Hoeyer, Klaus L., and Anja M. B. Jensen. "Transgressive Ethics: Professional Work Ethics as a Perspective on 'Aggressive Organ Harvesting.'" *Social Studies of Science* 43, no. 4 (2012): 598–618.

Hsu, Elizabeth. *The Transmission of Chinese Medicine*. Cambridge, UK: Cambridge University Press, 1999.

HUGO Pan-Asian SNP Consortium. "Mapping Human Genetic Diversity in Asia." *Science* 326 no. 5959 (December 9, 2009): 1541–1545.

Jablonka, E., and M. J. Lamb. *Evolution in Four Dimensions: Genetic, Epigenetic, Behavioral, and Symbolic Variation in the History of Life*. Cambridge, MA: MIT Press, 2006.

Jain, Lochlan. *Malignant: How Cancer Becomes Us*. Berkeley: University of California Press, 2013.

Jasanoff, Sheila. *Designs on Nature: Science and Democracy in the United States and Europe*. Princeton: Princeton University Press, 2007.

Jinan, T. A., L. C. Hong, M. E. Phipps, M. Stoneking, M. Ameem, and J. Edo. "Evolutionary History of Continental Southeast Asians: 'Early Train' Hypothesis Based on Genetic Analysis of Mitochondrial and Autosomal DNA Data." *Molecular Biology and Evolution* 29, no. 11 (November 2012): 3513–3527.

Jorde, L. B. and S. P. Wooding. "Genetic Variation, Classification and 'Race.'" *Nature Genetics* 36 (2004): S28–S33.

Kaufman, Sharon R. "Construction and Practice of Medical Responsibility: Dilemmas and Narratives from Geriatrics." *Culture, Medicine, and Psychiatry* 21 (1997): 1–12.

Kay, Lilly E. *The Molecular Vision of Life*. Oxford: Oxford University Press, 1993.

Keating, Peter, and Albert Cambrosio, *Biomedical Platforms*. Cambridge, MA: MIT Press, 2003.

Keck, Frederic. "Hong Kong as a Sentinel Post." *Limn* no. 3 (2013). http://limn.it/preface-sentinel-devices-2/.

Keck, Frederic, and Andrew Lakoff. "Preface: Figures of Warning." *Limn* no. 3 (2013). http://limn.it/preface-sentinel-devices-2/.

Kelty, C. M. "Preface." In *Clouds and Crowds*, edited by C. M. Kelty, L. Irani, and N. Seaver. *Limn* no. 2 (2012): 4–9.

Kirksey, S. Eben, and Stefan Helmreich. "The Emergence of Multispecies Ethnography." *Cultural Anthropology* 25, no. 4 (2010): 545–576.

Kleinman, Arthur. *Patients and Healers in the Context of Culture.* Berkeley: University of California Press, 1981.

———. *The Healing Narratives: Suffering, Healing and the Human Condition.* New York: Basic Books, 1989.

Kowal, Emma. "Orphan DNA: Indigenous Samples, Ethical Biovalue, and Postcolonial Science." *Social Studies of Science* 43, no. 4 (2013): 557–597.

Kuhn, Thomas S. *The Structure of Scientific Revolutions.* Chicago: University of Chicago Press, 1962.

Kundera, Milan. *Milan Kundera: The Art of the Novel.* Translated by Linda Ascher. New York: Grove, 1988.

Lakoff, Andrew. "Diagnostic Liquidity: Mental Illness and the Global Trade in DNA." *Theory and Society* 34, no. 1 (2005): 63–69.

———. "The Generic Biothreat, or, How We Became Unprepared." *Cultural Anthropology* 23, no. 3 (2008): 399–438.

———. *Pharmaceutical Reason: Knowledge and Value in Global Psychiatry.* Cambridge, UK: Cambridge University Press, 2006.

———. "Preparing for the Next Emergency." *Public Culture* 19, no. 2 (2007): 247–271.

Lakoff, Andrew, and Stephen J. Collier, eds. *Biosecurity Interventions: Global Health and Security in Question.* New York: Columbia University Press, 2008.

Larson, Christina. "Inside China's Genome Factory." *MIT Technology Review* (March/April 2013).

Latour, Bruno. *Reassembling the Social: An Introduction to Actor-Network Theory.* Oxford: Oxford University Press, 2005.

———. *Science in Action: How to Follow Scientists and Engineers through Society.* Cambridge, MA: Harvard University Press, 1987.

———. "Visualisation and Cognition: Drawing Things Together." *Knowledge and Society: Studies in the Sociology of Culture and Present* 6 (1986): 1–40.

———. "Why Has Critique Run Out of Steam? From Matters of Fact to Matters of Concern." *Critical Inquiry* (Winter 2004): 225–248.

Latour, Bruno, and Steven Woolgar. *Laboratory Life: The Construction of Scientific Facts.* Beverly Hills, CA: SAGE, 1979.

Law, John, and Annemarie Mol. "Situating Technoscience: An Inquiry into Spatialities." *Society and Space* 19 (2001): 609–621.

Lewis, M. W., and K. E. Wigen. *The Myth of Continents: A Critique of Metageography.* Berkeley: University of California Press, 1997.

Lim, M. K. "Transforming Singapore Health Care: Public-Private Partnership." *Annals Academy of Medicine Singapore* 34, no. 7 (2005): 461–467.

Litzinger, Ralph A. *Other Chinas: The Yao and the Politics of National Belonging.* Durham, NC: Duke University Press, 2000.

Liu, Jennifer A. "Making Taiwanese (Stem Cells): Identity, Genetics, and Hybridity." In *Asian Biotech: Ethics and Communities of Fate,* edited by Aihwa Ong and Nancy N. Chen, 239–262. Durham, NC: Duke University Press, 2010.

Lock, Margaret. *Encounters with Aging: Mythologies of Menopause in Japan and North America.* Berkeley: University of California Press, 1993.

————. *Twice Dead: Organ Transplants and the Reinvention of Death*. Berkeley: University of California Press, 2002.

Lock, Margaret, and Vinh-Kim Nguyen. *The Anthropology of Biomedicine*. Oxford: Wiley-Blackwell, 2010.

Lowe, Celia. "Viral Clouds: Becoming H5N1 in Indonesia." *Cultural Anthropology* 25, no. 4 (2010): 625–649.

Luhmann, Niklas. *Observations on Modernity*. Stanford, CA: Stanford University Press, 1998.

MacIntyre, Alasdair. *After Virtue: A Study in Moral Theory*. Third edition. Notre Dame, IN: University of Notre Dame Press, 2007.

Malinowski, Bronislaw. *Argnonauts of the Western Pacific*. New York: E. P. Dutton, 1922.

Marcus, George E. "The End(s) of Ethnography: Socio/Cultural Anthropology's Signature Form of Producing of Knowledge in Transition." *Cultural Anthropology* 23, no. 1 (February 2008): 1–14.

Marks, Jonathan. "Commentary: Toward an Anthropology of Genetics." *American Anthropologist* 116, no. 4 (2014): 749–751.

Martin, Emily. *Flexible Bodies: The Role of Immunity in American Culture from the Days of Polio to the Age of AIDS*. New York: Beacon, 1995.

Marx, Karl. *Capital: A Critique of Political Economy*. Chicago: Charles H. Kerr, 1906. Original published 1867.

Mauss, Marcel. *The Gift: Forms and Functions of Exchange in Archaic Societies*. Eastford, CT: Martino Fine Books, 2011. Original published 1954.

Mintz, Sydney W. *Sweetness and Power: The Place of Sugar in Modern History*. London: Penguin, 1986.

Mitchell, Timothy. *Rule of Experts: Egypt, Techno-Politics, Modernity*. Berkeley: University of California Press, 2002.

Mitchell, W., C. Breen, and M. Enzeroth. "Genomics as Knowledge Enterprise: Implementing an Electronic Research Habitat at the Biopolis Experimental Therapeutics Center." *Biotechnology Journal* 3, no. 3 (2008): 364–369.

Miyasaki, Hiro. *The Method of Hope*. Stanford, CA: Stanford University Press, 2004.

Mok, Timothy, et al. "Gefitinib or Carboplatin–Paclitaxel in Pulmonary Adenocarcinoma." *New England Journal of Medicine* 361 (September 2009): 947–957.

Mol, Annemarie. *The Logic of Care: Health and the Problem of Patient Choice*. New York: Routledge, 2008.

Montoya, Michael. *Making the Mexican Diabetic: Race, Science, and the Genetics of Inequality*. Berkeley: University of California Press, 2011.

Moor, Liz, and Celia Lury. "Making and Measuring Value." *Journal of Cultural Economy* 4, no. 4 (2011): 439–454.

Morton, Timothy. *Hyperobjects: Philosophy and Ecology after the End of the World*. Minneapolis: University of Minnesota Press, 2013.

Mountain, J. L., and N. Risch. "Assessing Genetic Contributions to Phenotypic Differences among 'Racial' and 'Ethnic' Groups." *Nature Genetics Supplement* 36, no. 11 (November 2004): 548–553.

Mukherjee, Siddhartha. *The Emperor of All Maladies: A Biography of Cancer*. London: Fourth Estate, 2011.

Mullaney, Thomas S. *Coming to Terms with the Nation: Ethnic Classification in Modern China*. Berkeley: University of California Press, 2010.

Nelson, Alondra. "DNA Ethnicity as Black Social Action?" *Cultural Anthropology* 28, no. 3 (2013): 527–536.

Neyrat, Frederic. "The Birth of Immunopolitic." Translated by A. De Boever. *Parrhesia* 10 (2010): 31–38.

Nguyen, Vinh-Kim. "Antiretroviral Globalism, Biopolitics, and Therapeutic Citizenship." In *Global Assemblage: Technology, Politics and Ethics as Anthropological Problems*, edited by Aihwa Ong and Stephen J. Collier, 124–144. Malden, MA: Wiley-Blackwell, 2005.

———. *The Republic of Therapy: Triage and Sovereignty in West Africa's Time of AIDS*. Durham, NC: Duke University Press, 2010.

Nonini, Donald M. *"Getting By": Class and State Formation among Chinese in Malaysia*. Ithaca, NY: Cornell University Press, 2015.

Normille, D. "Consortium Hopes to Map Human History in Asia." *Science* 306, no. 5702 (December 3, 2004): 166.

———. "SNPS Study Supports Southern Migration Route to Asia." *Science* 326, no. 5959 (December 11, 2009): 1470.

Olds, Kris, and Nigel Thrift. "Cultures on the Brink: Reengineering the Soul of Capitalism—on a Global Scale." In *Global Assemblage: Technology, Politics and Ethics as Anthropological Problems*, edited by Aihwa Ong and Stephen J. Collier, 270–290. Malden, MA: Wiley-Blackwell, 2005.

The 1000 Genomes Project Consortium. "An Integrated Map of Genetic Variation from 1,092 Human Genomes." *Nature* 491 (November 1, 2012): 56–65. http://www .nature.com/nature/journal/v491/n7422/full/nature11632.html.

Ong, Aihwa. "Assembling around SARS: Technology, Body Heat, and Political Fever in Risk Society." In *Ulrich Beck: Kosmopolitisches Projekt*, edited by A. Pferl and N. Szaider, 81–88. Baden-Baden: Nomos Verlagsgesellschaft, 2004.

———. "A Biocartography: Maids, Neoslavery, and NGOs." In *Neoliberalism as Exception*, 195–217. Durham, NC: Duke University Press, 2006.

———. "Ecologies of Expertise: Assembling Flows, Managing Citizenship." In *Global Assemblage: Technology, Politics and Ethics as Anthropological Problems*, edited by Aihwa Ong and Stephen J. Collier, 337–351. Malden, MA: Wiley-Blackwell, 2005.

———. *Flexible Citizenship: The Cultural Logics of Transnationality*. Durham, NC: Duke University Press, 1999.

———. "Introduction: An Analytics of Biotechnology and Ethics at Multiple Scales." In *Asian Biotech: Ethics and Communities of Fate*, edited by Aihwa Ong and Nancy N. Chen, 1–51. Durham, NC: Duke University Press, 2010.

———. "Lifelines: The Ethics of Banking Blood for Family and Beyond." In *Asian Biotech: Ethics and Communities of Fate*, edited by Aihwa Ong and Nancy N. Chen, 190–214. Durham, NC: Duke University Press, 2010.

———. "A Milieu of Mutations: The Pluripotency and Fungibility of Life in Asia." *East Asian Science, Technology and Society* 7 (2013): 1–18.

———. *Neoliberalism as Exception: Mutations in Citizenship and Sovereignty*. Durham, NC: Duke University Press, 2006.

———. "Please Stay: Pied-a-Terre Subjects in the Global City." *Citizenship Studies* 11, no. 1 (2007): 83–93.

———. "Scales of Exception: Experiments with Knowledge and Sheer Life in Tropical Southeast Asia." *Singapore Journal of Tropical Geography* 29, no. 2 (2008): 1–13.

———. *Spirits of Resistance and Capitalist Discipline: Factory Women in Malaysia.* Albany: State University of New York Press, 1987.

———. "Zoning Technologies in East Asia." In *Neoliberalism as Exception,* 97–120. Durham, NC: Duke University Press, 2006.

Ong, Aihwa, and Nancy N. Chen, eds. *Asian Biotech: Ethics and Communities of Fate.* Durham, NC: Duke University Press, 2010.

Palsson, Gisli, and Paul Rabinow. "The Icelandic Controversy: Reflections on the Transnational Market of Civic Virtue." In *Global Assemblage: Technology, Politics and Ethics as Anthropological Problems,* edited by Aihwa Ong and Stephen J. Collier, 91–104. Malden, MA: Wiley-Blackwell, 2005.

Petryna, Adryna. "Ethical Variability: Drug Development and Globalizing Clinical Trials." *American Ethnologist* 32, no. 2 (2005): 182–197.

———. *When Experiments Travel: Clinical Trials and the Global Search for Human Subjects.* Princeton: Princeton University Press, 2009.

Pollock, Anne. *Medicating Race: Heart Disease and Durable Preoccupations with Difference.* Durham, NC: Duke University Press, 2012.

Preston, Richard. *The Hot Zone: A Terrifying True Story.* New York: Random House, 1995.

Quammen, David. *Spillover: Animal Infections and the Next Human Pandemic.* New York: W. W. Norton, 2012.

Rabinow, Paul. "Artificiality and Enlightenment: From Sociobiology to Biosociality." In *Essays on the Anthropology of Reason,* 91–110. Princeton: Princeton University Press, 1996.

———. *French DNA: Trouble in Purgatory.* Chicago: University of Chicago Press, 1999.

———. *French Modern: Norms and Forms of the Social Environment.* Cambridge, MA: MIT Press, 1989.

Rabinow, Paul, and Gaymon Bennett. *Designing Human Practices: An Experiment with Synthetic Biology.* Chicago: University of Chicago Press, 2012.

———. "Synthetic Biology: Ethical Ramifications 2009." *Systems and Synthetic Biology* 3, nos. 1–4 (2009): 99–108.

Raffles, Hugh. *Insectopedia.* New York: Pantheon, 2010.

Rapp, Rayna. "Moral Pioneers: Women, Men, and Fetuses on a Frontier of Reproductive Technology." In *Embryos, Ethics, and Women's Rights,* edited by E. D. Baruch, A. F. D'Adamo and J. Seager, 101–116. New York: Haworth, 1988.

Rawls, John. *A Theory of Justice.* Cambridge, MA: Belknap Press of Harvard University Press, 1999.

Reardon, Jenny. *Race to the Finish: Identity and Governance in an Age of Genomics.* Princeton: Princeton University Press, 2005.

Redfield, Peter. *Life in Crisis: The Ethical Journey of Doctors without Borders.* Berkeley: University of California Press, 2013.

Rheinberger, Hans-Jorg. *An Epistemology of the Concrete: Twentieth-Century Histories of Life.* Durham, NC: Duke University Press, 2010.

Rose, Nikolas. *The Politics of Life Itself.* Princeton: Princeton University Press, 2007.

Rose, Nikolas, and Carlos Novas. "Biological Citizenship." In *Global Assemblage: Technology, Politics and Ethics as Anthropological Problems,* edited by Aihwa Ong and Stephen J. Collier, 439–463. Malden, MA: Wiley-Blackwell, 2005.

Samimian-Darash, Limor. "Governing Future Potential Biothreats: Toward an Anthropology of Uncertainty." *Current Anthropology* 54, no. 1 (2013): 1–22.

Schein, Louisa. *Minority Rules: The Miao and the Feminine in China's Cultural Politics.* Durham, NC: Duke University Press, 2000.

Scheper-Hughes, Nancy. "The Last Commodity: Post-human Ethics and the Global Traffic in 'Fresh' Organs." In *Global Assemblage: Technology, Politics and Ethics as Anthropological Problems,* edited by Aihwa Ong and Stephen J. Collier, 145–167. Malden, MA: Blackwell, 2005.

Schneider, David. *A Critique of the Study of Kinship.* Ann Arbor: University of Michigan Press, 1984.

Shapin, Steven. *The Scientific Life: A Moral History of a Late Modern Vocation.* Chicago: University of Chicago Press, 2008.

Shouse, E. "Feeling, Emotion, Affect." *M/C Journal* 8, no. 6 (2005). http://journal.media-culture.org.au/0512/03-shouse.php. Accessed March 14, 2013.

Skloot, Rebecca. *The Immortal Life of Henrietta Lacks.* New York: Random House, 2010.

Sleeboom-Faulkner, Margaret. "Embryo Controversies and Governing Stem Cell Research in Japan." In *Asian Biotech: Ethics and Communities of Fate,* edited by Aihwa Ong and Nancy N. Chen, 215–236. Durham, NC: Duke University Press, 2010.

———. *Global Morality and Life Science Practices in Asia: Assemblages of Life.* London: Palgrave Macmillan, 2014.

Specter, Michael. "The Gene Factory: Letter from Shenzhen." *New Yorker,* January 6, 2014, 34–43.

Star, S. L., and J. R. Greisemer. "Institutional Ecology, 'Translation,' and Boundary Objects: Amateurs and Professionals in Berkeley's Museum of Vertebrate Zoology, 1907–39." *Social Studies of Science* 19, no. 3 (1989): 387–420.

Strathern, Marilyn, ed. *Audit Cultures.* London: Routledge, 2000.

Sunder Rajan, Kaushik. *Biocapital: The Constitution of Postgenomic Life.* Durham, NC: Duke University Press, 2006.

———. "The Experimental Machinery of Global Clinical Trials: Case Studies from India." In *Asian Biotech: Ethics and Communities of Fate,* edited by Aihwa Ong and Nancy N. Chen, 55–80. Durham, NC: Duke University Press, 2010.

———. *Lively Capital: Biotechnologies, Ethics, and Governance in Global Markets.* Durham, NC: Duke University Press, 2012.

Sung, Wen-ching. "Chinese DNA: Genomics and Bionation." In *Asian Biotech: Ethics and Communities of Fate,* edited by Aihwa Ong and Nancy N. Chen, 263–292. Durham, NC: Duke University Press, 2010.

Talbot, Marianne. *Bioethics: An Introduction.* Cambridge, UK: Cambridge University Press, 2012.

Taleb, Nassim Nicholas. *The Black Swan: The Impact of the Highly Improbable.* Second edition. London: Penguin, 2010.

Taussig, Michael T. "Reification and the Consciousness of the Patient." *Social Science and Medicine* 148 (2010): 3–13.

Thompson, Charis. "Asian Regeneration? Nationalism and Internationalism in Stem Cell Research in South Korea and Singapore." In *Asian Biotech: Ethics and Communities of Fate*, edited by Aihwa Ong and Nancy N. Chen, 91–117. Durham, NC: Duke University Press, 2010.

———. *Good Science: The Ethical Choreography of Stem Cell Research*. Cambridge, MA: MIT Press, 2013.

———. *Making Parents: The Ontological Choreography of Reproductive Technologies*. Cambridge, MA: MIT Press, 2005.

———. "Race Science," *Theory, Culture and Society* 23, nos. 2–3 (2006): 547–549.

Traweek, Sharon. *Beamtimes and Lifetimes: The World of High Energy Physicists*. Cambridge, MA: Harvard University Press, 1992.

Unschuld, P. U. *Medicine in China: A History of Ideas*. Berkeley: University of California Press, 1985.

van Hoyweghen, Inez. "On the Politics of Calculative Devices: Performing Life Insurance Markets." *Journal of Cultural Economy* 7, no. 3 (2014): 334–352.

———. "Taming the Wild Life of Genes by Law? Genes Reconfiguring Solidarity in Private Insurance." *New Genetics and Society* 29, no. 4 (December 2010): 431–455.

Vasantkumar, Christopher. "Han at Minzu's Edges: What Critical Han Studies Can Learn from China's 'Little Tibet.'" In *Critical Han Studies*, edited by Thomas S. Mullaney, James Leibold, Stephane Gros, and Eric Vanden, 234–256. Berkeley: University of California Press, 2012.

Vora, Kalinda. *Life Support: Biocapital and the New History of Outsourced Labor*. Minneapolis: University of Minnesota Press, 2015.

Wade, Nicholas. "East Asian Physical Traits Linked to 35,000-Year-Old Mutation." *New York Times*, February 15, 2013.

Wade, P., C. L. Beltran, E. Restrepo, and R. V. Santos, eds. *Mestizo Genomics: Race Mixture, Nation, and Science in Latin America*. Durham, NC: Duke University Press, 2014.

Wailoo, Keith. *How Cancer Crosses the Color Line*. Oxford: Oxford University Press, 2011.

Waldby, Catherine. "Biobanking in Singapore: Post-developmental State, Experimental Population." *New Genetics and Society* 28, no. 3 (2009): 253–265.

———. "Stem Cells, Tissue Cultures, and the Production of Biovalue." *Health: An Interdisciplinary Journal for the Social Study of Health, Illness and Medicine* 6, no. 3 (2002): 305–323.

Waldby, Catherine, and Robert Mitchell. *Tissue Economies: Blood, Organs and Cell Lines in Late Capitalism*. Durham, NC: Duke University Press, 2007.

Wallerstein, Immanuel. *The Modern World-System, Vol. I: Capitalist Agriculture and the Origins of the European World-Economy in the Sixteenth Century*. New York: Academic Press, 1974.

Washington, Henrietta. *Medical Apartheid: The Dark History of Medical Experimentation on Black Americans from Colonial Times to the Present*. New York: Doubleday, 2007.

Weber, Max. *On Charisma and Institution Building*. Edited by S. N. Eisenstadt. Chicago: University of Chicago Press, 1968.

———. "Science as a Vocation." In *From Max Weber: Essays in Sociology*, edited and translated by H. H. Gerth and C. Wright Mills, 129–156. New York: Oxford University Press, 1946. Original published 1922.

Whitington, Jerome. "Fingerprinting, Bellwether, Model Event: Climate Change as Speculative Anthropology." *Anthropology Theory* 13, no. 4 (2012): 308–328.

Wilson, Ara. "Medical Tourism in Thailand." In *Asian Biotech: Ethics and Communities of Fate*, edited by Aihwa Ong and Nancy N. Chen, 118–143. Durham, NC: Duke University Press, 2010.

Wolfe, Nathan. *The Viral Storm: The Dawn of a New Pandemic Age*. New York: William Holt, 2011.

Yarborough, M. A. "Increasing Enrollment in Drug Trials: The Need for Greater Transparency about the Social Value of Research in Recruitment Efforts." *Mayo Clinic Proceedings* 88, no. 5 (2013): 442–445.

Zhan, Mei. *Other-Worldly: Making Chinese Medicine through Transnational Frames*. Durham, NC: Duke University Press, 2009.

Zhou, Hongyan, Shili Wu, Jin Young Joo, Saiyong Zhu, Dong Wook Han, Tongxiang Lin, Sunia Trauger, Geoffery Bien, Susan Yao, Yong Zhu, Gary Siuzdak, Hans R. Schöler, Lingxun Duan, and Sheng Ding. "Generation of Induced Pluripotent Stem Cells Using Recombinant Proteins." *Cell Stem Cell* 4 (May 8, 2009): 381–384.

Žižek, Slavoj. "Philosophy, the 'Known Unknowns,' and the Public Use of Reason." *Topoi* 25, nos. 1–2 (2006): 137–142.

Asian genetics research (*continued*)
medicine and, 145–47; racialized genomics in, 31–32; scientific entrepreneuralism
in, 8–11; Singapore as hub for, 105–12;
stem cell research and, 139–43, 147–53;
zoonotic in infectious disease and, 179–81
"Asian tiger" economies: biomedical science
challenges in, 114–17; Singapore's designation as, 3–4, 9
Asia-Pacific Economic Cooperation
(APEC), 164
Association of Southeast Asian Nations
(ASEAN), 192
"athlete gene" research, 197–99, 213–22
authoritarianism: BGI in context of, 199–203;
disease management and, 188–94; genomics research in, 37–39, 42–43, 48–49;
tissue and organ banks and role of, 57–59
autism, cognitive genetics and, 231
auto-immunology, cancer genetics and,
110–11
autopoiesis, 137–39

Backdoor to Eugenics (Duster), xv–xvi
bar-code model of ethnicity, 43–47, 62
Barnett, Robert, 215
Bateson, Gregory, xvii, 15–16
Beck, Ulrich, 20
Berlant, Lauren, 78
best practices: Asian skepticism concerning,
97–98; at BGI, 234–35; at Biopolis, 105;
in cancer research, 108; cosmopolitan
science and, 22–23; epidemiology and
intervention, 191; ethical capitalism and,
93; ethnicity in genetics and, 111; infectious disease research, 192–93; intra-Asian
collaboration on, 48
BGI Genomics, xx, xxii; DNA research at,
17–18, 26; ethnic classification and, 207–13;
ethnic heuristic and, 227–28; founding
and organizational structure of, 199–203;
future challenges for, 224–26, 235–37;
goals of, 197–99; impact on government
policy of, 230–32; PRC genomic analog,
218–22; scientific entrepreneurialism in,
11, 233–34; securities research at, 24–25;
Tibetan-HAN DNA study and, 213–17; tree
of life models, 203–7

biodiversity. *See* genetic diversity
bioethics: in Asian genomics research,
52–54; BGI research and, 210–13; in
cancer research, 107–9; center-periphery
bioethics, 94–96; cognitive genetics and,
231; in DNA research, 98–105; human and
animal testing and, 59–60, 247n16; human
embryonic research and, 141–43; in
multiethnic societies, 54–56; organ transplants and, 56–59; problematization of,
96–98; productive uncertainty of, 93–112;
scientific entrepreneurship and, 134–35;
in stem cell research, 139–41, 150–53;
uncertainty in biomedical research and,
23; value for native donors and, 103–5
Bioethics Advisory Committee (BAC), 54,
143, 225–26
biology: genomic science and, xvi–xxii; as
storytelling, 160
biomarkers: for Asian cancer patients, 81–82,
111–12; female Asian nonsmoker cancer
patient biomarker, 82–86
biomedical assemblage: biosecurity and,
194–96; infectious disease research and,
180–81, 190–94; science in context of,
xiii–xv, 236–37
biomedical platform, construction of, 22,
105–12
Biomedical Research Council, 38–39
biomedical sciences: Asian research in,
xii–xv; biopiracy and, 52–54; Chinese
ethnic heuristic in, 45–47; ecosystem
approach to, 18–19; kinship systems and,
74–75; Singapore's initiatives in, 5
Biopolis life-sciences hub, x–xii, xiv–xv;
Asian genetic research at, 47–49, 62–65,
88–89; autopoiesis research at, 137–39;
BGI and, 201–3, 207–13; bioethics at, 23,
94–112; cancer genetics at, 65–66, 88–89,
107–9; Chinese cancer cartography and,
67–72; development and organization
of, 1–3, 240n2, 242n26; disease research
at, 52–54, 62–65; DNA research in,
14–18, 225–26; ecosystem configuration
of, 18–19; epidemiological research at,
25–26; ethical norms at, 22, 105–12; ethnic
heuristic in, 45–47, 229–30; foreign lab
workers at, 126–29; future challenges for,

224–26, 235–37; genetic variants collection initiative, 39–41; intra-Asian rivalry at, 147–50; local science talent at, 130–34; pan-Asianism in, 161–64; public-private partnerships and, 38–39; race/ethnicity issues in research of, xix–xxii; scientific entrepreneurialism in, 8–11, 22; scientific virtue in, 116–17; securities research in, 24–25; situated cosmopolitan science in, 7–8; stem cell research at, 141–47, 151–53; superstar scientists at, 117–25; sustainability issues for, 62–65, 232–33; transethnic genomics and, 32–35, 34–35, 49–50; uncertainty in research at, 20–26; viral-ethnic fold in disease research and, 182–83, 236–37

biopolitics: of deferral, 78–81; global pandemics and, 188–94; of power, 12, 20; scientific entrepreneurialism and, 232–33

biosentinels, in infectious disease, 177–78

biosociality, 36; genetic exceptionalism and, 80–81

biosovereignty, xiii, 11, 60–62

biostatistics: ethnicity and, 43–45; transethnic genomics and, 34–35; "wild" genes and, 35–39

biovalue: definition of, 14–15; in DNA research, 99–105; tissue donor ethics and, 143–45

blood donation, Chinese genomics and, 212–17

Bongsu, Arif (Dr.), 54, 138, 148

boundary objects theory: "biosovereignty" and, xiii, 11; cancer biomarkers and, 81–82; cancer treatment and, 74–77, 86–87

Bourdieu, Pierre, 53–54

Bowker, Geoffrey, 17–18, 253n10

Brahmachari, Samir, 168

British medical training, xxi, 132–134. *See also* United Kingdom, Singaporean scientists training in

BRCA1 gene, 75

breakbone fever, 182–83

breast cancer: Asian biomarkers for, 82; current research on, 73–75

Brenner, Sydney (Dr.), 5, 114–15, 241n9

Briggs, Charles, 94–95

Bring Your Genes to Cal program, ix–x

Broad Institute (Harvard and MIT), 200

Buchanan, Brian, 81

Buddhism, bioethics and, 54–56

Bumrungrad International Hospital, 62

Bush, George W., 142

California Institute for Regenerative Medicine, 142–43

Cambridge University, 209

Cambrosio, Albert, 22

Camus, Albert, 188–90

cancer research: affect and, 77–78; Asian biomarkers in, 81–82; Asian genes and, 62–65; Asian-oriented goals in, 87–89; BGI and, 208–13; biopolitics of deferral and, 78–81; Chinese cancer cartography, 67–72; Chinese ethnic heuristic in, 46–47; drug development in, 245n21; family history and, 73–75; genetic diversity and, 75–77; "one-step solution" in, 107–9; in Singapore, 48–49; taxonomy and scalability in, 65–66

capitalism: bioethics and, 52–54; center-periphery bioethics and, 95–96; medical technology and, 22

cascade model, infectious disease research, 178–79

Catholicism, bioethics and, 54–56

Cavalli-Sforza, Luca, 157–58, 250n19

Celera Genomics, 9, 30–32

cell fate modulation, 136–37, 236–37

center-periphery bioethics, 94–96, 211–13

Centers for Disease Control (CDC), 174–77, 184–85; SARS super-spread phenomena and, 188–90

Chen, Nancy, 11

Chen (Dr.) (pseudonym), 201, 212, 217–21, 229–30

China. *See* People's Republic of China

Chinese Academy of Medical Sciences, 199, 209

Chinese ethnic heuristic, 45–47

Chua (Dr.) (pseudonym), 57–58

Chua Beng Huat, 4, 32–33

CIMO (Chinese, Indian, Malay-Muslim, Others), xviii; ethnic framework (Singapore) and, 32–35 civic virtue, scientific research and, 23–24, 40, 42–43, 114, 131–34

climate change, evovability and combinability genetics and, 185–90

clinical trials: bioethics and, 59–60, 94–96, 105–12; "one-step solution" in, 107–9; outsourcing of, 48–49; pan-Asian clinical trial, Iressa cancer therapy, 83–86; Singapore resources for, 21, 105–12

cloning: intra-Asian rivalry on, 148; Singapore legislation on, 54–56; stem cell research and, 59

co-ethnicity: Chinese cancer cartography and, 70–72; in DNA databases, 36–39

cognitive genetics, 230–31

Cohen, Lawrence, xii

Cold Spring Harbor Laboratory, 76

Collier, Stephen J., xx, 20, 115

Colman, Alan, 2, 55, 64, 118, 142–45

colonialism: Chinese ethnic taxonomy and, 208–13; genomics and, xviii; illness and disease imagery linked to, 113–17; impact on Singapore of, 13–14; infectious disease research and, 174–77; medical science resistance to, 113–17; multiracial government and, 32–35; social Darwinism and, 247n1

combinability, in infectious disease research, 185–90

The Coming Plague, 177–78

Complete Genomics, 200–201

computational technology, genomics research and, 37–39

Confucianism, 33

Contagion, 177–78

contingency analysis, risk genomics and, 22

contract research organizations (CROs): bioethics and, 95–98; cancer research and, 48–49

contrary affects, cancer research and, 77–78

Cooper, Melinda, 144–45

cosmopolitan science: autopoiesis and, 137–39; BGI impact on, 199–203; Biopolis initiatives in, 5; future challenges for, 224–26, 235–37; pluralization of, 72, 232–35; scientific virtue in, 115–17; as situated science, 7–8

Crispr method of gene editing, 233, 254n14

cross-border collaboration: biosovereignty issues and, 61–62; Chinese cancer cartography and, 67–72; disease emergence research, 24–25; DNA data and, 13–14

culture: bioethics and, 54–56; biomarkers, 82–86; Chinese genomics and, 209–13; CIMO ethnic divisions and, 33–35; cultural sensitivity in clinical trials, 108–9; ethnicity and, 43–45; gene-culture interaction and, 44–45; informed consent from native donors and, 103–5; population genetics and, 230

customized therapy, Chinese cancer cartography and, 70–72

database management: anonymized data, 42–43; Asian genomics research and, 35–39; BGI DNA databases, 204–7; bioethics and, 108–9; Chinese cancer cartography and, 68–72; Chinese ethnic heuristic and, 46–47; genetic diversity and, 40–41; plug and play platform for, 49; self-identified ethnicity data in, 43–45

data diversity, defined, 17–18

Dawkins, Richard, 161

Dayak people, 103

deCODE Genetics database, 32, 39–41

Defense Science Organization, 38–39

deferral, biopolitics of, 78–81

degenerative diseases, stem cell research and, 145–47

Deleuze, Gilles, 16, 77–78

dengue fever: diagnosis and management of, 182–83; molecular mechanisms in, 184–85; urban surveillance programs, 183–85

Development Bank of China, 200

difference, in scientific research, xvi–xxii

differentiation, in scientific research, xvi–xxii, 225–26

disease atlases, 21; Asian disease research and, 51–52, 62–65; Chinese cancer cartography, 67–72

disease emergence research, 24–25, 30; anonymized data and, 42–43; Asian genetics and, 62–65, 232–33; biopolitics and, 190–94; biosecurity assemblage and, 194–96; Chinese ethnic heuristic in, 46–47; evovability and combinability genetics in, 185–90; folding scales of intervention in, 182–85, 236–37; induced pluripotent stem cell technology and, 153; kinship systems and, 74–75; NEIDs (newly emerging infectious diseases), 177–78; tissue and organ banks and, 57–59; tropical diseases, 174–77, 184–85; urban surveillance approaches in, 183–85; zoonotic rezoning in, 179–81

DNA research: anonymized data and, 42–43; Asian DNA databases, 21, 159–60; BGI dominance in, 197–207, 210–13; bioethics in, 98–105; biosovereignty issues and, 61–62; in China, 26; database management in, 35–39; ethnic heuristic in, 45–47; fortune-cookie genetics and, ix–xii; fungibility in, 14–18; pan-Asian approach to, 158–61, 169, 172–73; race and gender and, xvi; single wave hypothesis in, 167–68; switching station narrative in, 169–72

DNA sequencing, 201–7, 247n18

donor ethics, 103–5, 143–45, 273n27

double-body technology, 145–47, 153

Doudna, Jennifer, 233

Dr. No (film), 219–20

Duara, Prasenjit, 163–64

Duke-NUS Graduate Medical School, xxi–xxii; Biopolis life-sciences hub collaboration and, 18–19, 105–12, 118–25; Chinese cancer cartography and, 71–72; foreign lab workers at, 126–29; "global schoolhouse" initiative and, 131–34; infectious disease research and, 184–85; pandemic research by, 25–26; programs in Singapore, 119–20; tropical disease research at, 174–77; viral-ethnic fold in disease research and, 182–83

Dumit, Joseph, 80–81

Duster, Troy, xv–xvi

"East Asian" biomarker, 83–86

East Asian Science, Technology and Society, xiii

Ebola virus, 179, 190, 195–96

ecology: genomics and, xvii; of ignorance, 19–20; in infectious disease research, 181

Economic Development Board (Singapore), 4

ecosystems: BGI tree of life model, 204–7; in biomedical research, 18–19, 51–54; zoonotic in infectious disease and, 179–81

educational system, CIMO ethnic divisions and inequality in, 33–35

electronic health records: anonymized health data and, 42–43; in China, 48

Eli Lilly, 208

embryonic stem cells: bioethics and, 141–43; donor ethics and, 143–45

Emory University, 195

English language, Asian genomics research and use of, 47

Enigma machine, xi–xii

environment, genomic science and, xvi–xxii

EPAS1 gene, 214–17

epidemiological research, 25–26, 174–77; Singapore as focus of, 180–81, 190–94

epidermal growth factor receptor (EGFR): Asian biomarkers and, 82; somatic mutation in Asian patients, 83–86

epigenetics: biomarkers and, 81–82; Chinese ethnic classifications and, 209–13; gene-culture interaction and, 44–45

epistemic object, bioethics and, 100–103, 247n20

Erwin, Kathleen, 212

ES Cell International (ESI), 145

ethical variability concept, 95–96

Ethnic Classification Project (*minzu shibie*), 208–13

ethnic heuristic: Asian biomarkers and, 82; BGI and, 207–13, 227–28; bioethics and, 100–103; cancer cartography and, 67–72; in China, 45–47, 67–72, 208–13; gene-culture interactions and, 44–45; mobility and, 49–50; transethnic genomics and, 32–35; tribalism and, 58–59

ethnicity: Asian biomarkers and, 82; barcode model of, 43–47; in BGI research, 207–17; biosociality and, 80–81; cancer biomarkers and, 84–86; disease risk and, 63–65, 232–33; geography and, 69–72; in medical research, xv–xvi, 17–18, 232–37; in pan-Asian genomics research, 165–67; in Singapore's genomic research, 12–14,

geography: ethnicity and, 69–72; genomics and, xvii

Germany, biomedical research in, 124–25, 134–35

germ-line cells, biomarkers and, 81–82

gift economy concept, DNA research ethics and, 99–105

GlaxoSmithKline, 193–94

Gleevec, 75

"global assemblage," xiii, 7

globalization: Asian recruitment in manufacturing and, 149–50; BGI projects and, 203; human and animal testing and, 60; infectious disease research and, 179–81; pharmaceutical industry and, 110–12; transethnic genomics and, 33–34; zoonotic in infectious disease and, 179–81

"global schoolhouse" initiative, 131–34

"governing through blood" program, 212–13

"Greater Asian" (*Da Yaxiyazhuyi*), 163–64

"Greater East Asia Co-prosperity Sphere," 250n11

The Great Influenza, 177–78

Greenhalgh, Susan, 230–31

Guyer, Jane, 12

H5N1 viral specimen controversy, 6

Hadid, Zaha, 1

Han (graduate student) (pseudonym), 127–29

Han population: Chinese cancer cartography and, 67–72; Chinese DNA research on, 26, 208–13. *See also* Tibetan-Han DNA study

Hapmap gene catalog, 39–41

Haraway, Donna, 72, 137, 160

healthcare services: affect and, 77–78; anonymized data and, 42–43; public-private partnerships in, 37–39

Heart of Darkness, 177–78

Helix Bridge (Singapore), 223–27

hepatitis B virus, 208

Herceptin, 75

high altitude performance, genetic studies of, 217–18

Hinduism, bioethics and, 54–56

Hispanic patients, US cancer clinical trials for, 84–86

Hoeyer, Klaus L., 97

homogeneity, genetic diversity research and limits of, 40–41

hope, analytics of, 78–81, 246n14

hot spots, in infectious disease, 177–78

The Hot Zone, 177–78

Human Genome Organization (HUGO), 8, 61–62, 241n21; Asian genetics research and, 158, 164–67, 172–73; Hapmap gene catalog and, 39, 165, 250n13; pan-Asianism and, 171–72

Human Genome Project, xi, 30–32, 157–58, 164, 166; BGI Genomics and, 199

Human Organ Transplant Act (HOTA), 56–59

Hwang Woo Suk, 148

Iceland, genomics research in, 32, 39–41

Illumina Company, 200

imagined communities: Chinese genomics and, 211–12; tissue and organ donation and, 58–59

The Immortal Life of Henrietta Lacks (Skloot), 273n27

immunology: autopoiesis and, 137–39; biopolitics and, 233–34; cancer genetics and, 110–11; cell fate modulation and, 137; dengue fever research and, 184–85; genetics and, 64–65; stem cell research and, 145–47, 153

immutable mobile: ethnicity as, 21, 36–37, 43–47; infectious disease research and, 178–79

imperialism. *See* colonialism

India: ethnicity in genomics research and, 45–47; generic drug development in, 247n9; lab workers from, 126–29; population genetics in, 168; scientific entrepreneuralism in, 9

indigenous communities: center-periphery bioethics and, 94–96; DNA research and, 99–105, 157–60

Indonesia: bioethics and, 143–46, 152–53; H5N1 viral specimen controversy, 6; immunology and, 145–46; induced pluripotent stem (iPS) cell technology and, 24; inter-Asian rivalry and, 147–50;

research, 116–17; cancer treatment and, 74–77, 86–87; funding reliability issues with, 122–25; in genomics research, 37–39; infectious disease research and, 182–83; protein chip technology and, 124–25; scientific entrepreneurialism and, 9–11

quality metrics, scientific entrepreneuralism and, 107–9
Quan (Dr.) (pseudonym), 55
Qing (graduate student) (pseudonym), 127–29

Rabinow, Paul, 36, 40, 52
race: anonymized health data and, 43; in BGI research, 208–13; genomic science and, xvi–xxii, 17–18, 160–61; in medical research, xv–xvi; in pan-Asian genomics research, 165–67
racialized genomes, 31–32
Ranbaxy, 96, 247n9
Rawls, John, 54–56
Reagan, Ronald, 144
Reardon, Jenny, 160
reassortment of genes, infectious disease research, 185–90
reciprocal exchange, disease mapping and, 53–54
recombinant DNA technology, 144
Reeve, Christopher, 144
Regional Emergency Disease Initiative (REDI) Center, 191–94
regional genetic research: BGI collaboration in, 201–3; genetic diversity and, 41; national jurisdiction and, 103–5; pan-Asianism and, 161–64; role of expatriates in, 114–17
religion: bioethics and, 54–56; human embryonic research and, 141–43; organ transplants and, 56–59
Rheinberger, Hans-Jörg, 7
risk genomics, 20–21; anonymized health data and, 42–43; autopoiesis and, 137–39; cancer prevalence in China and Asia and, 21–22; ethnicity and disease and, 62–65, 232–33
risk society, genomics and, 20

Rizal, José (Dr.), 113
RNA virus sequencing: dengue fever research and, 184–85; viral-ethnic fold in disease research and, 182–83
Rose, Nikolas, 12
"rule of experts," colonial legacy of, 33
Rumsfeld, Donald, 24, 220, 243n53

SARS (severe acute respiratory syndrome) outbreak, 4–5, 10–11, 121–25; anonymized data from, 42–43; Chinese susceptibility to, 66; evolvability and combinability genetics in, 185–90; infectious disease research and, 175–77, 179–81; pan-Asianism and, 164; predictive technologies for, 178; super-spread phenomenon, 188–90
scale: Chinese science and, 203, 212, 216, 219, 235; elasticity of, 191–92; ethnic indicator and, 36, 44–47, 57–58, 66; genomic research and, 139, 161; of intervention, 181–82, 195; of problematization, 19–20, 25–26, 178–81; scientific progress and, 235–36; temporality of, 12; as zone of action, 170–76
Schein, Louisa, 211
Schell, Orville, 3
Science magazine, 168
scientific entrepreneuralism: in Asia, 8–11; biomedical platforms and, 22, 105–12; contrary affects and, 77–78; "one-step solution" in trials and, 107–9; superstar scientists and, 117–25; sustainability of, 62–65, 232–33; uncertainties and, 22
scientific research: affect and, 77–78; in Asia, xii–xv; in China, 197–99; ethnography of, 45; as storytelling, 160–61; as vocation, 114–17
scientific virtue: conception of, 113–17; scientific entrepreneurship and, 134–35; superstar scientists and, 117–25; unregulated science and, 234–35
security: anthropological research on, 20; Chinese genetics research and concerns over, 200–203; disease as threat to, 159–60, 177–79, 186–90; global assemblage and, 194–96; research initiatives in, 24–25

Security, Population, and Territory
(Foucault), 20, 79
self-curing body paradigm: stem cell
research and, 145–47, 153; traditional
Chinese medicine (TCM) and,
226–27
Sequoia Capital, 202
Shapin, Steven, xiii-xiv, 9, 11, 114–15, 172
Sheng Ding, 149
Singapore: biomedical science challenges
in, xviii–xxi, 114–17; biosecurity assemblage in, 194–96; cancer therapy
research in, 75–77; clinical testing and
medical tourism in, 21, 105–12; demographics and economics of, 3–7; disease
surveillance in, 183–90; ethnic genetics
research in, 21, 29–30; expatriate
scientists in, 118–25; infectious disease
research in, 121–25, 174–77, 180–81,
190–94; local science talent in, 130–34;
migrant population in, 121–25; pan-
Asianism and role of, 162–64; pluripotency in research by, 12–14; "risk-free"
zone for scientific research in, 22–23;
stem cell bank development in, 150;
as switching station, 169–72, 190–94;
transethnic genomics research in,
32–35. *See also* Biopolis life-sciences hub
Singapore Genome Variation Project
(SGVP), 31–32
Singapore Immunology Network (SIgN),
184–85
Singapore Medicine initiative, 244n20
Singapore-MIT Alliance for Research
and Technology (SMART), 105–12,
124–25
Singapore National Science Council, 5
Singapore Novartis Institute for Tropical
Diseases, 184–85, 194
single nucleotide polymorphisms (SNPs),
39–41, 215–17
single wave hypothesis, 167–68
situated cosmopolitan science, 7–8
situated virtue, 23–24, 40–43, 113–17
social Darwinism, 247n1
somatic cell nuclear transfer (SCNT),
148–49

somatic mutation, 82; Asian cancers and,
82–86
Sommers (Dr.) (pseudonym), 106, 124–25,
134–35
source area development, infectious disease
research and, 191–94
South China founder effect, 82
South Korea: pan-Asian research and,
169; scientific research in, 106; stem cell
research in, 148–49
Spillover, 177–78
Spinoza, Baruch, 77–78
state-controlled research: anonymized data
and, 42–43; in Asia, 17–18; BGI Genomics and, 199–203, 219–22; bioethics and,
96–98; biotechnology and, 52–54; in
China, 208–13; foreign lab worker recruitment by, 126–29; local scientists recruitment and, 130–34; stem cell research and,
141–43
stem cell research, 119–20, 136–37; autopoiesis and, 137–39; cancer therapy and,
64–65; ethics of, 24, 139–41, 150–53;
future challenges in, 150–52; intra-Asian
rivalry in, 147–50; patient-specific stem
cells, 145–47; in Singapore, 54–56; transitions in, 139–41
storytelling, in pan-Asian genetics research,
160–61, 164–67
stratified medicine, development of, 14
Sunder Rajan, Kaushik, xii, 95–96
Sun Yat-Sen (Dr.), 113, 163–65
Sun Yat-Sen University, 233
superstar scientists, 117–25
Svensen (Dr.) (pseudonym), 166, 201
switching station narrative of Singapore,
169–72
Synthetic Genomics, 30–32

Tai (Dr.) (pseudonym), 47–49
Talbot, Marianne, 97
Tan, W. (Dr.), 64–65, 71–72, 93–94,
110–112
Tan Chorh Chuan (Dr.), 252n33
targeted cancer therapy: BGI Genomics
and, 207–13; ethnicity and, 75–77, 82–86;
expanded access to, 84–86